T0310664

LONDON MATHEMATICAL SOCIETY LECTURE NOTE SERIES

Managing Editor: Professor J.W.S. Cassels, Department of Pure Mathematics and Mathematical Statistics, University of Cambridge, 16 Mill Lane, Cambridge CB2 1SB, England

The titles below are available from booksellers, or, in case of difficulty, from Cambridge University Press.

London Mathematical Society Lecture Note Series. 218

Surveys in Combinatorics, 1995

Edited by

Peter Rowlinson
University of Stirling

CAMBRIDGE
UNIVERSITY PRESS

CAMBRIDGE UNIVERSITY PRESS
Cambridge, New York, Melbourne, Madrid, Cape Town, Singapore, São Paulo

Cambridge University Press
The Edinburgh Building, Cambridge CB2 8RU, UK

Published in the United States of America by Cambridge University Press, New York

www.cambridge.org
Information on this title: www.cambridge.org/9780521497978

© Cambridge University Press 1995

First published 1995

A catalogue record for this publication is available from the British Library

ISBN 978-0-521-49797-8 paperback

Transferred to digital printing 2008

Contents

Preface

The British Combinatorial Conference returns to Scotland in 1995 with the fifteenth in a series of international meetings concerned with all branches of Combinatorics. Nine distinguished researchers, representing both Mathematics and Computing Science, have been invited to deliver the principal lectures, and this volume contains the survey articles which they have submitted in advance. These contributions certainly justify the early interest shown in the conference, and should pave the way towards another successful meeting. The second essential ingredient is the series of short talks presented by delegates outwith the plenary sessions, and papers contributed in conjunction with these will be considered for a special edition of *Discrete Mathematics* edited by Douglas Woodall.

The year 1995 marks the hundredth anniversary of the death of the Rev.T.P.Kirkman, and one day of the conference will be designated 'Kirkman Day' in his honour. The talks on this day, including those by Rosa and Spence, will be devoted to topics related to Kirkman's achievements in Combinatorics.

I am grateful to the authors and referees for their co-operation in meeting the necessary deadlines, and I am indebted to Roger Astley of Cambridge University Press for his assistance in the preparation of the text. The British Combinatorial Committee acknowledges with thanks the financial support provided by the London Mathematical Society and the Institute of Combinatorics and its Applications.

Peter Rowlinson
Stirling, February 1995.

EUCLIDEAN GEOMETRY OF DISTANCE REGULAR GRAPHS

C. D. Godsil [1]

Combinatorics and Optimization
University of Waterloo
Waterloo, Ontario
Canada N2L 3G1

ABSTRACT

A graph is distance regular if it is connected and, given any two vertices u and v at distance i, the number of vertices x at distance j from u and k from v is determined by the triple (i, j, k). Distance regular graphs are interesting because of their connections with coding theory, design theory and finite geometry.

We can introduce geometric methods into the study of distance regular graphs as follows. Let X be a graph on n vertices and identify the i-th vertex of X with the i-th standard basis vector e_i in $\mathbf{R}^n$. Let θ be an eigenvalue of (the adjacency matrix of) X and let U be the corresponding eigenspace. We then associate to the i-th vertex of X the image of e_i under orthogonal projection onto U. If U has dimension m then we have a mapping from $V(X)$ into $\mathbf{R}^n$. If X is distance regular, then it can be shown that the image of $V(X)$ lies in a sphere centred at the origin, and that the cosine of the angle between the vectors representing two vertices u and v is determined by the distance between them in X. In this paper we survey some of the applications of these methods.

[1] Support from grant OGP0093041 of the National Sciences and Engineering Council of Canada is gratefully acknowledged.

1. Introduction

We define a *representation* of a graph G in $\mathbf{R}^m$ to be a map, ρ, from $V(G)$ into $\mathbf{R}^m$, such that for any two vertices u and v, the inner product

$$\langle \rho(u), \rho(v) \rangle$$

is determined by the distance $\text{dist}(u, v)$ between u and v in G. Because any vertex is at distance 0 from itself, this implies that the image of $V(G)$ under ρ lies on a sphere centred at the origin. We will always assume that ρ is not the zero map, although it might map all vertices of G to the same non-zero vector.

To obtain a non-trivial example, consider the dodecahedron embedded as a regular polytope in $\mathbf{R}^3$ with centre of mass at the origin. We may assume that its vertices lie on the unit sphere. If G is the 1-skeleton of this polytope then the map that assigns to each vertex of G the vector representing it in $\mathbf{R}^3$ is a representation. Further examples can be obtained from the other Platonic solids.

Representations provide us with the opportunity to apply geometric methods to graph theory, in particular to the study of distance-regular graphs.

2. Representations

We first introduce one concept from linear algebra. The *Gram matrix* of a set of vectors $v_1, \ldots, v_n$ is the $n \times n$ matrix M with $M_{i,j}$ equal to $\langle v_i, v_j \rangle$. It is not hard to see that a Gram matrix is positive semi-definite and symmetric. The converse is also true, any symmetric positive semi-definite matrix is a Gram matrix.

If G is a graph with diameter d, let G_i denote the graph with the same vertex set as G, with two vertices adjacent in G_i if and only if they are at distance i in G. Let A_i denote the adjacency matrix of G_i, with the understanding that $A_0 = I$. We call $A_0, \ldots, A_d$ the *distance matrices* of G. If ρ is a representation of G in $\mathbf{R}^m$, then the matrix $M(\rho)$ has rows and columns indexed by the vertices of G and, if $u, v \in V(G)$, then

$$(M(\rho))_{u,v} := \langle \rho(u), \rho(v) \rangle.$$

The matrix $M(\rho)$ is a linear combination of the matrices A_i. It is a Gram matrix, of the vectors ρ for u in $B(G)$, and so it is positive semi-definite. As every symmetric positive semi-definite matrix is a Gram matrix, we see conversely that each positive semi-definite matrix in the span of $A_0, \ldots, A_d$ determines a representation of G.

A subgraph H of G is *isometric* if $\mathrm{dist}_H(u, v) = \mathrm{dist}_G(u, v)$ for any two vertices u and v of H. It follows that if ρ is a representation of G then its restriction to any isometric subgraph H is a representation of H.

We have not required that a representation be injective. We can sometimes show that if $\mathrm{dist}(u, v) \leq r$, then $\rho(u) \neq \rho(v)$; if this is the case we will say that ρ is *r-injective*.

By way of introduction, we offer the following.

Lemma 2.1. *There are only finitely many graphs G such that both G and its complement admit an injective representation into $\mathbf{R}^m$.*

Proof. If ρ is injective, then the image of any clique is a regular simplex. Hence if G has a representation in $\mathbf{R}^m$ then there are no cliques in G with more than $m + 1$ vertices in them. Because the complement $\overline{G}$ of G also has a representation on $\mathbf{R}^m$, there can be no independent set in G with size greater than $m + 1$. Our claim follows by Ramsey's theorem. □

Lemma 2.2. *If G admits a 2-injective representation into $\mathbf{R}^m$, then the maximum valency of a vertex in G is bounded by a function of m.*

Proof. If N is the neighourhood of a vertex in G, then two distinct vertices in N are at distance 1 or 2 in G. Hence if ρ is a 2-injective representation of G into $\mathbf{R}^m$, then the restriction of ρ to N is a representation of both N and $\overline{N}$ in an $(m - 1)$-dimensional affine subspace of $\mathbf{R}^m$. So, by the previous lemma, $|V(N)|$ is bounded by a function of m. □

If we compute an explicit expression for the function in this lemma using Ramsey theory, we obtain an exponential bound for the maximum valency of G. We will see in the next section that this is far too large.

3. Geometry

As the image of a graph under a representation is a set of points in $\mathbf{R}^m$, we need some terminology for such subsets. Let S be a subset of the unit sphere in $\mathbf{R}^m$. The *degree* of S is the size of its *degree set*

$$\{\langle x, y \rangle : x, y \in S, \ x \neq y\}.$$

A set with degree s is called an *s-distance set*. If f is a polynomial in m variables then f is a function on S. We say that S is a *spherical t-design* if the average over the points of S of any polynomial f with degree at most t is equal to the average value of f over the unit sphere. This terminology is based on an analogy with the theory of t-designs; roughly speaking a spherical t-design is a finite approximation to the unit sphere, whereas a t-(v, k, λ) design is an approximation to the set of all k-subsets of a fixed set of v elements. For more on this viewpoint, see [7: Chapter 14]. For information on spherical designs see [4] and [7: Chapter 13]. The largest integer t such that S is a t-design is the *strength* of S. A subset S of the unit sphere is a 1-design if and only if the sum of the vectors in it is the zero vector.

The following bound is a combination of results from [4], in particular Theorems 2.4 and 5.11.

Theorem 3.1. *Let S be a subset of the unit sphere in $\mathbf{R}^m$ with degree s and strength t, and define $f(r)$ to be $\binom{m+r-1}{r} + \binom{m+r-2}{r-1}$. Then*

$$f(\lfloor t/2 \rfloor) \leq |S| \leq f(s);$$

further, if one of these inequalities is an equality, then so is the other.

A subset S for which equality holds in the bound of this theorem is called a *tight* spherical design. It implies that a 2-distance set in $\mathbf{R}^m$ has size at most $m(m+3)/2$, and therefore this value may be taken as the bound of Lemma 2.2.

Theorem 3.2. *Let S be a subset of the unit sphere in $\mathbf{R}^m$ with degree s and strength t. If $x \in S$ and α belongs to the degree set of S then $\{y \in S : \langle x, y \rangle = \alpha\}$ is a $(t + 1 - s)$-design.*

Let S be a finite set of points on the unit sphere in $\mathbf{R}^m$. If α lies in the degree set of S, let A_α be the $(0, 1)$-matrix with rows and columns indexed by S, and with $(A_\alpha)_{u,v} = 1$ if and only if $\langle u, v \rangle = \alpha$.

Theorem 3.3. *Let S be a subset of the unit sphere in $\mathbf{R}^m$ with degree s and strength t, and let Δ be its degree set. If $t \geq 2s - 2$, then the matrices A_α for α in Δ, together with the identity matrix, form an s-class association scheme.*

A finite set of symmetric $(0, 1)$-matrices $A_0, \ldots, A_d$ forms an association scheme if:

(a) $A_0 = I$,
(b) $\sum_\alpha A_\alpha = J$,
(c) for all i and j, the product $A_i A_j$ is a linear combination of $A_0, \ldots, A_d$.

For more on association schemes see, e.g., [3, 7]. It is not too hard to show that $t \leq 2s$ for any finite subset of the unit sphere, so the lower bound on t in this theorem is quite strong. An important consequence of the last theorem is that a 2-distance set with strength at least 2 gives rise to a complementary pair of strongly regular graphs. Further, all strongly regular graphs can be obtained from such subsets of the unit sphere.

4. Distance Regular Graphs

A graph G with diameter d is *distance regular* if, for $i = 1, \ldots, d$, the distance matrix A_i is a polynomial in A_1 with degree i. In more combinatorial terms, G has the property that if integers i, j and k are given and u and v are vertices in G at distance i then the number of vertices in G at distance j from u and distance k from v is independent of the choice of u and v. A graph is *distance transitive* if, given any two ordered pairs of vertices (u, v) and (u', v') such that $\mathrm{dist}(u, v) = \mathrm{dist}(u', v')$, there is an automorphism of

G that maps (u, v) to (u', v'). It is easy to see that a distance transitive graph is distance regular. The 1-skeletons of the Platonic solids in $\mathbf{R}^3$ are all distance-transitive; this is an almost immediate consequence of the fact that they are regular polytopes. For a complete introduction to the theory of distance regular (and distance transitive) graphs, see [3].

We point out that there are vast numbers of distance regular graphs that have trivial automorphism group; distance regularity does not imply the existence of any non-identity automorphisms in general.

The *Johnson graph* $J(v, \ell)$ has all ℓ-subsets of a fixed set V of size v as its vertices, with two ℓ-subsets adjacent if they intersect in exactly $\ell - 1$ points. This is a distance transitive graph, and therefore distance regular. Let Q be a fixed set of size q. The graph $J(v, 2)$ is also known as the line graph of the complete graph. The *Hamming graph* $H(n, q)$ has vertex set Q^n, and two elements of Q^n are adjacent if they differ in exactly one coordinate. (The set Q is often taken to be a finite field, but we do not even require it to have prime power order.) The Hamming graph $H(n, 2)$ is n-cube. The Hamming graphs are all distance transitive.

The distance matrices of a distance regular graph with diameter d form a basis of a commutative algebra of dimension $d + 1$ over the reals, called the *Bose-Mesner algebra*. Thus, if ρ is a representation of G, then $M(\rho)$ is a positive semi-definite matrix in the Bose-Mesner algebra of G; conversely every such matrix gives rise to a representation of G.

One class of positive semi-definite matrices can be obtained using projections. Let θ be an eigenvalue of $A = A_1$ with multiplicity m, let U be the corresponding eigenspace and let E_θ be the matrix representing orthogonal projection on U. Then E_θ can be shown to be a polynomial in A, and therefore it lies in the Bose-Mesner algebra of G. Hence we obtain a representation of G in $\mathbf{R}^m$, which we will call an *eigenspace representation* of G.

As E_θ lies in the span of the matrices $A_0, \ldots, A_d$, its diagonal entries are all equal. Because $E_\theta^2 = E_\theta$, all eigenvalues of E_θ are equal to 0 or 1; therefore $\operatorname{tr} E_\theta = \operatorname{rk} E_\theta$ and each diagonal entry of E_θ is equal to $m/|V(G)|$.

So there are constants $w_0, \ldots, w_d$, with $w_0 = 1$, such that

$$\frac{|V(G)|}{m} E_\theta = \sum_{i=0}^{d} w_i A_i. \tag{4.1}$$

As this matrix is positive semi-definite, each principal 2×2 submatrix has non-negative determinant, which implies that

$$|w_i| \leq 1.$$

We can view w_i as the cosine of the angle between $\rho(u)$ and $\rho(v)$, for any two vertices u and v at distance i in G. We call $w_0, \ldots, w_d$ the *sequence of cosines* of G; this sequence depends on the eigenvalue θ. We can summarise our conclusions as follows.

Lemma 4.1. *Let G be a distance regular graph with diameter d, let θ be an eigenvalue of G, with cosine sequence $w_0, \ldots, w_d$ and let ρ be the corresponding representation of G. Then*

$$M(\rho) = \sum_{i=0}^{d} w_i A_i. \qquad \square$$

If $A = A_1$ then $AM(\rho) = \theta M(\rho)$. As A is a $(0, 1)$-matrix, this implies that

$$\theta \rho(u) = \sum_{v \sim u} \rho(v). \tag{4.2}$$

(We write $v \sim u$ to denote that v is adjacent to u.) We also have the following.

Lemma 4.2. *Let G be a distance regular graph with valency k and let θ be a non-trivial eigenvalue of G. If ρ is the representation of G on the eigenspace belonging to θ, then $\rho(G)$ is a spherical 2-design.* $\square$

If G is a distance regular graph with valency k, then k is an eigenvalue of G. The associated eigenspace is spanned by $\mathbf{1}$, the vector with all entries equal to 1, and hence the representation we obtain is not interesting. Further, $-k$ is an eigenvalue of G if and only if G is bipartite; the associated eigenspace is, once again, 1-dimensional. We will call an eigenvalue θ of G non-trivial if $\theta \neq \pm k$.

5. Cosines

If G is distance regular, then there are constants a_i, b_i and c_i such that

$$AA_i = b_{i-1}A_{i-1} + a_iA_i + c_{i+1}A_{i+1}. \tag{5.1}$$

(This is essentially the matrix formulation of the combinatorial definition of distance regularity given at the start of the previous section.) Here $a_0 = c_0 = 0$, $c_1 = 1$ and $b_d = 0$. We denote the valency of G by k, so $b_0 = k$. Further, a_1 is the number of triangles on an edge of G and c_2 is the number of common neighbours of two vertices at distance 2 in G. By way of example, for the Johnson graph $J(v, k)$ we have

$$b_i = (k-i)(v-k-i), \quad a_i = i(v-2i), \quad c_i = i^2,$$

when $i = 0, \ldots, k$. In all cases we have that

$$b_i + a_i + c_i = k, \qquad i = 0, \ldots, d.$$

If x is at distance i from u in the distance regular graph G then there are b_i neighbours of u at distance $i+1$ from x, a_i at distance i and c_i at distance $i-1$. Hence, if ρ is the representation of G on the eigenspace belonging to θ and we take the inner product of both sides of (4.2) with $\rho(x)$, then we obtain:

$$\theta w_i = c_i w_{i-1} + a_i w_i + b_i w_{i+1}. \tag{5.2}$$

As $w_0 = 1$, this implies that $\theta = b_0 w_1$ and thus:

$$w_1 = \frac{\theta}{k}. \tag{5.3}$$

Further, $\theta w_1 = 1 + a_1 w_1 + b_1 w_2$ and $1 + a_1 + b_1 = k$, whence

$$w_2 = \frac{(\theta - a_1)w_1 - 1}{b_1} = \frac{\theta^2 - a_1\theta - k}{k(k-1-a_1)}. \tag{5.4}$$

We see that if $w_1 = 1$, then $\theta = k$ and the corresponding eigenspace is spanned by the vector $\mathbf{1}$. If $w_1 = -1$ then $\theta = -k$. In this case G must be bipartite, the multiplicity of θ is 1 and the corresponding eigenvector takes value 1 on one colour class of G and -1 on the other. The next result answers most questions about injectivity.

Lemma 5.1. *Let G be a connected distance regular graph with diameter d and valency k, where $k \geq 3$. Let $\theta_0 > \theta_1 > \cdots > \theta_d$ be the eigenvalues of G. The eigenspace representation corresponding to θ_i is not injective if and only if*

(a) *$i = 0$, (that is, $\theta = k$), or*
(b) *$\theta = -k$ and G is bipartite, or*
(c) *i is even and G is antipodal.* ☐

A distance regular graph G is said to be *antipodal* if G_d is not connected, in which case the components of G_d all have the same size. Complete multipartite graphs are precisely the antipodal distance regular graphs with diameter 2. The proof of Lemma 5.1 given on [7: p. 265] also yields the following.

Lemma 5.2. *Let G be a distance regular graph with valency k, and let θ be a non-trivial eigenvalue of G. If G is not complete multipartite then the representation belonging to θ is 2-injective.* ☐

The number of *sign-changes* in a sequence $a_0, \ldots, a_d$ is the number of indices i such that $a_i a_{i+1} < 0$. Deleting the terms of a sequence that are equal to 0 does not change the number of sign-changes in it. A proof of the next result appears in [7: Lemma 13.3.1].

Lemma 5.3. *Let G be a distance regular graph with diameter d and let θ be the i-th largest eigenvalue of G. Then the cosine sequence $w_0, \ldots, w_d$ has exactly i sign-changes, and the sequence $w_0 - w_1, \ldots, w_{d-1} - w_d$ has exactly $i - 1$ sign-changes.* ☐

From (5.4) above we see that

$$1 - w_2 = \frac{k - \theta}{k} \frac{\theta - a_1 + k}{k - a_1 - 1}. \tag{5.5}$$

As $|w_2| \leq 1$, one consequence of this is that

$$a_1 - k \leq \theta,$$

for any eigenvalue θ of G. We also have

$$1 - 2w_1 + w_2 = \frac{k - \theta}{k} \frac{k - a_1 - 2 - \theta}{k - a_1 - 1}. \tag{5.6}$$

The parameter $1 - 2w_1 + w_2$ turns out to be important; partial evidence for this is provided by the next result, taken from [3: Prop. 4.4.9].

Theorem 5.4. *Let G be a distance regular graph with diameter d and valency k, let θ be a non-trivial eigenvalue of G and let $w_0, \ldots, w_d$ be the corresponding sequence of cosines. If G contains an induced copy of C_4, then $1 - 2w_1 + w_2 \geq 0$. If equality holds, then $\theta = k - a_1 - 2$ and $w_i = 1 - i(a_1 + 2)/k$.*

Proof. Let ρ be the representation of G on the eigenspace belonging to θ and let u, v, x and y be the images under ρ of the vertices in an induced 4-cycle in G; where u and x represent vertices at distance 2 in G, as do v and y. Then

$$0 \leq \|u + x - v - y\|^2 = 4(1 - 2w_1 + w_2) \tag{5.7}$$

proving our claim.

Suppose equality holds in (5.7), and suppose that z is the image under ρ of a vertex at distance $i - 1$ from u and distance $i + 1$ from x. Then $u + x - v - y = 0$ and therefore

$$0 = \langle z, u + x - v - y \rangle = w_{i-1} + w_{i+1} - 2w_i.$$

Given this it is not hard to show that $w_i = 1 - i\frac{a_1 + 2}{k}$. $\qquad\square$

Lemma (Delsarte) 5.5. *Let G be a distance regular graph with diameter d and valency k. If C is a clique in G of size c, then $c \leq 1 - (k/\theta_d)$.*

Proof. If θ is an eigenvalue of G, then the $c \times c$ matrix

$$I + w_1(J - I)$$

is positive semi-definite, because it is a positive constant times a principal submatrix of the positive semi-definite matrix E_θ. Hence all its eigenvalues are non-negative. As each row of $I + w_1(J - I)$ sums to $1 + (c - 1)w_1$, it follows that

$$0 \leq 1 + (c - 1)\frac{\theta}{k}.$$

This bound is strongest when θ is as small as possible, so take it to be θ_d. $\square$

As a final application, consider a distance regular graph G with diameter 3. From (5.4) and (5.2) we obtain

$$w_1 - w_2 = \frac{k - \theta}{k} \frac{\theta + 1}{k - 1 - a_1}. \tag{5.8}$$

Suppose that $\theta = -1$ and that ρ is the corresponding representation. As $w_1 = w_2$, it follows that $\rho(G)$ is a 2-distance set and, by Lemma 4.2, it is a spherical 2-design. Now Theorem 3.3 implies G_3 must be strongly regular. An example is provided by the Johnson graph $J(7, 3)$.

6. Multiplicity Bounds

There are a number of results which can be viewed as providing a lower bound on the multiplicity of an eigenvalue of a distance regular graph.

The first result can be motivated by the following observation. Suppose that G is a distance regular graph with valency k and that $\theta \neq k$ is an eigenvalue of G with multiplicity m. If G has no triangles, then the image of the neighbours of a vertex under the representation ρ associated to θ is a regular simplex lying on an affine hyperplane in $\mathbf{R}^m$, whence it follows that $k \leq m$.

If the girth of G is at least $2m$, then the set of vertices at distance less than $\lfloor m/2 \rfloor$ from a fixed vertex of G is an isometric subgraph of G, and a tree. Terwilliger [21] used this to prove a result that implies the following.

Theorem (Terwilliger) 6.1. *Let G be a distance regular graph with valency k and girth g. Then any non-trivial eigenvalue of G has multiplicity at least $k^{\lfloor g/4 \rfloor}$.* $\square$

We now present a version of some results from Terwilliger [23]. If $u \in V(G)$, let $G_i(u)$ denote the set of vertices in G at distance i from u. If

ρ is a representation of a distance regular graph G in $\mathbf{R}^m$ and $u \in V(G)$, then, as we noted earlier, the image under ρ of $G_1(u)$ is a 2-distance set in $\mathbf{R}^{m-1}$. Further, if B is the adjacency matrix of the subgraph of G induced by $G_1(u)$, then $\rho(G_1(u))$ has Gram matrix

$$I + w_1 B + w_2(J - I - B) = (1 - w_2)\left(I + \frac{w_1 - w_2}{1 - w_2}B + \frac{w_2}{1 - w_2}J\right).$$

As this matrix is positive semi-definite, any eigenvalue of it corresponding to an eigenvector orthogonal to $\mathbf{1}$ must be non-negative. If τ is such an eigenvalue, then

$$\frac{w_1 - w_2}{1 - w_2}\tau \geq -1. \tag{6.1}$$

If the points in $\rho(G_1(u))$ are linearly dependent then our Gram matrix above must be singular, and so equality must hold in (6.1) for some eigenvalue τ (with $\tau \neq k$). Note also that if G is distance regular then $G_1(u)$ is regular with valency a_1, and so the eigenvectors for any eigenvalue of $G_1(u)$ other than a_1 is orthogonal to $\mathbf{1}$. Using Equations (5.5) and (5.8), we see that if equality holds in (6.1), then

$$\tau = -1 - \frac{b_1}{\theta + 1}. \tag{6.2}$$

We can summarise our observations as follows.

Theorem 6.2. *Let G be a distance regular graph with diameter d and let θ be an eigenvalue of G with cosine sequence $w_0, \ldots, w_d$. If $u \in V(G)$ and τ is an eigenvalue of $G_i(u)$ then*

$$\frac{w_1 - w_2}{1 - w_2}\tau \geq -1,$$

with equality if and only if $\rho(G_1(u))$ is linearly dependent. $\square$

Terwilliger [23] also shows that, if equality holds, then θ is θ_1 or θ_d and either it is an integer that divides $k - a_1 - 1$, or θ_1 and θ_d are the zeros of an irreducible quadratic polynomial over the integers.

It is clear that if the valency k of G is greater than the multiplicity m of θ then $\rho(G_1(u))$ must be linearly dependent. But this can happen in other cases. It is not hard to show that, if $a_d = 0$, then the vectors

$$\rho(x) - \rho(y), \qquad (x, y \in G_d(u))$$

are orthogonal to the vectors in $\rho(G_1(u))$. Thus if

$$k + \dim(G_d(u)) - 1 > m,$$

then $\rho(G_1(u))$ is linearly dependent. For applications of this, see [9].

Our last result for this section concerns the diameter.

Theorem ([5]) 6.3. *Let G be a distance regular graph with diameter d and let θ be a non-trivial eigenvalue of G with multiplicity m, where $m > 2$. Then $d \leq 3m - 4$, with equality if and only if G is the 1-skeleton of the dodecahedron.* □

We outline the proof of this. The key step is to show that $b_{m-1} = 1$. Let u and v be two vertices at distance $m - 1$ in G, and let P be a path of length $m - 1$ joining them. Then P is isometric. Suppose that z and z' are two neighbours of v in G at distance m from u. Then, for each vertex x on P we have

$$\mathrm{dist}(x, z) = \mathrm{dist}(x, z')$$

and therefore

$$\langle \rho(x), \rho(z) \rangle = \langle \rho(x), \rho(z') \rangle.$$

This implies that $\rho(z) = \rho(z')$ and, as ρ is 2-injective and $\mathrm{dist}(z, z')) = 2$, that $z = z'$. This shows that $b_{m-1} = 1$; the remainder of the argument is graph-theoretic, with no geometric content. The interested reader is referred to [5].

The assumption $m > 2$ is to exclude cycles, which have eigenvalues of multiplicity 2, and can have arbitrarily large diameter. Koolen [11: Theorem 7.17] has shown that the upper bound on d can be reduced to $2m - 1$.

Corollary 6.4. *If $m > 2$, then there are only finitely many distance regular graphs with an eigenvalue of multiplicity m that are not complete multipartite.*   □

These results raise the prospect of classifying the distance-regular graphs with an eigenvalue of small multiplicity. A reasonable amount of progress has been made on this task; see [15, 25, 26] for the graphs with an eigenvalue with multiplicity up to 7. Koolen [11: Section 7.3.3] has determined the graphs with an eigenvalue of multiplicity 8. No new distance regular graphs have appeared as a result of this classification, unfortunately.

7. Polytopes

Let G be distance regular with diameter d and with eigenvalues $\theta_0, \ldots, \theta_d$, in decreasing order. Let ρ be a representation of G in $\mathbf{R}^m$ corresponding to the eigenvalue θ. We call the convex hull of the points in the image of $\rho(G)$ an *eigenpolytope*. In this section we develop enough of the theory of convex polytopes to enable us to study these objects. Brøndsted's book [2] is a convenient reference for most of our polytopal needs. We assume some familiarity with the elements of the theory of convex sets.

A convex polytope is defined to be the convex hull of a finite set of points. The *dimension* of a polytope is the dimension of the smallest affine space which contains all its points; we will often refer to a polytope of dimension m as an *m-polytope*. A 0-polytope is a complicated name for a point.

An affine hyperplane H is a *supporting hyperplane* for a polytope $\mathcal{P}$ if it contains at least one point of $\mathcal{P}$, and if all points of $\mathcal{P}$ not on H lie on the same side of H. A *face* of $\mathcal{P}$ is any set of points $\mathcal{P} \cap H$, where H is a supporting hyperplane. Any face is itself a convex polytope, and a face of a face of $\mathcal{P}$ is a face of $\mathcal{P}$.

There is an alternative definition of faces which will be useful. Suppose that $\mathcal{P}$ is a polytope in $\mathbf{R}^m$. The set of points in $\mathcal{P}$ at which a linear functional on $\mathbf{R}^m$ takes its maximum value is a face, and all faces can be obtained in this way. Less formally, if $h \in \mathbf{R}^m$, then the points x in $\mathcal{P}$ such that $h^T x$ is maximal form a face of $\mathcal{P}$. It is not too hard to see that these two definitions of faces are equivalent.

An r-face is a face which has dimension r. A 0-face is usually called a *vertex* and a 1-face is called an *edge*. An $(m-1)$-face of an m-polytope is a *facet*. The vertices and edges of a polytope form a graph, which is the *1-skeleton* of the polytope.

Theorem 7.1. *If X is the 1-skeleton of an m-polytope $\mathcal{P}$ and C is a cutset in X, then the vertices in C span an affine hyperplane, and hence $|C| \geq m$.*

Proof. Let C be a subset of the vertices of X which does not span $\mathbf{R}^m$. If C is contained in a face of P, then $X \setminus C$ is connected by [2: Theorem 15.5]. Otherwise there is a hyperplane containing C and at least one other vertex of $\mathcal{P}$. The proof of Theorem 15.6 from [2] now yields that $X \setminus C$ is connected. $\square$

Balinski [1] proved that the 1-skeleton of an m-polytope is m-connected; this is proved in [2] as Theorem 15.6. Thus Theorem 7.1 is essentially a reformulation of this result, and we will also make use of it in this form. In either form, this result implies that the 1-skeleton of an m-polytope has minimum valency at least m.

A polytope is *simplicial* if every face is a simplex. An m-polytope is *simple* if every k-face lies in exactly $m - k$ facets. There is a more intuitive characterisation, given as Theorem 12.12 in [2].

Theorem 7.2. *An m-polytope is simple if and only if its 1-skeleton is regular of valency m.* $\square$

Theorem 7.3. *Let $\mathcal{P}$ be a simple polytope. Then:*
(a) Every face of $\mathcal{P}$ is simple.
(b) Suppose u and $v_1, \ldots, v_k$ are vertices of $\mathcal{P}$ such that uv_i is an edge of $\mathcal{P}$ for $i = 1, \ldots, k$ and let F be the smallest face of $\mathcal{P}$ containing u and the vertices v_i. Then F has dimension k and the edges uv_i are the only edges in F on u.

Proof. See Theorem 12.15 and 12.17 of [2]. $\square$

Suppose that A is the adjacency matrix of the graph G and θ is an eigenvalue of A with multiplicity m. Let $\mathcal{P}$ be the eigenpolytope of G belonging to θ and let h be a vector in $\mathbf{R}^m$. Then the function which maps u in V onto $\langle h, \rho(u) \rangle$ is an eigenvector of A with eigenvalue θ and each eigenvector of A with eigenvalue θ can be obtained in this way. As noted by Powers [19], the vertices on which an eigenvector assumes its maximum value form a face of $\mathcal{P}$. In particular if there is an eigenvector equal to 1 on u and v and less than 1 on all other vertices of G then uv is an edge of $\mathcal{P}$. We will make use of this later.

As also noted by Powers [19], equitable partitions can be used to derive information about the faces of eigenpolytopes. We explain this. If V is the vertex set of G and π is a partition of V, let $F(\pi)$ denote the vector space of all functions on V which are constant on the cells of π. Call π *equitable* if $F(\pi)$ is A-invariant. If π is an equitable partition of G then $F(\pi)$ contains eigenvectors for A, each of which must be constant on the cells of π. Therefore, at least two cells of π are faces of some eigenpolytope of G. (This is of course still true if we assume only that $F(\pi)$ contains an eigenvector of A, but I have found no use for this generality yet.) For an introduction to equitable partitions see [6: Section 5.1] and [7,10].

8. Hamming and Johnson Graphs

The Johnson graphs $J(v, \ell)$ and the Hamming graphs $H(n, q)$ have already been introduced. We note here that $H(n, 2)$ is the n-cube, that $J(v, 1)$ is the complete graph on v vertices and $J(v, 2)$ is the line graph of K_v. As $J(v, \ell)$ and $J(v, v-\ell)$ are isomorphic, it is convenient to assume that $v \geq 2\ell$. The Johnson graph $J(v, \ell)$ has valency $\ell(v-\ell)$, $a_1 = v-2$ and eigenvalues

$$\theta_i = \ell(v - \ell) - i(v + 1 - i), \qquad i = 0, \ldots, \ell.$$

In particular $\theta_1 = \ell(v - \ell) - v$ and

$$k - \theta_1 - a_1 + 2 = 0.$$

It follows from (5.6) that $1 - 2w_1 + w_2 = 0$ and, from Theorem 5.4, that

$$w_i = 1 - i\frac{v}{\ell(v - \ell)}, \qquad i = 0, \ldots, \ell. \tag{8.1}$$

Let ρ be the representation belonging to θ_1 and define a representation $\hat{\rho}$ by

$$\hat{\rho}(\alpha) = \sqrt{\frac{\ell(v-\ell)}{v}}\, \rho(\alpha).$$

Let D be the set of vectors

$$\{\hat{\rho}(\alpha) - \hat{\rho}(\beta)\}$$

obtained as α and β vary over the vertices of $J(v, \ell)$. Then (8.1) implies that the inner product of any two vectors in D is an integer.

Consequently the set of all integer combinations of vectors from D is an integral lattice, that can be shown to be generated by vectors

$$\{\hat{\rho}(\alpha) - \hat{\rho}(\beta)\},$$

where α and β are adjacent. As these vectors have length 2, it follows that it is a *root lattice*, and these can be classified. But note how little of the structure of $J(v, \ell)$ has been used to reach this point. Thus it is not surprising that this approach can be used to characterise the Johnson graphs. The same procedure works for the Hamming graphs and again yields a characterisation. This was first carried out by Terwilliger [22, 24] and Neumaier [17]. This work was extended in Chapter 3 of Brouwer et al. [3]; a revised and slightly more accurate version appears in Koolen [11].

9. Eigenpolytopes

The 1-skeletons of the octahedron, icosahedron and dodecahedron are all distance regular graphs and the θ_1-eigenpolytopes of these graphs are the corresponding regular solids. In other words, these graphs are isomorphic to the 1-skeletons of their θ_1-eigenpolytopes. This also holds for the complete graphs and n-cubes, and so we ask whether there are more examples.

Suppose that G is distance regular with diameter at least 2 and second largest eigenvalue θ_1. The cosine sequence for θ_1 is non-increasing, by Lemma 5.3. From (5.4) and (5.2) we have

$$w_1 - w_2 = \frac{(\theta_1 + 1)(\theta_1 - k)}{k}.$$

As G is not complete, $\theta_1 > -1$, and therefore the vertices v such that $\|\rho(u) - \rho(v)\|$ is minimal are precisely the neighbours of u in G. Because the points in $\rho(G)$ lie on a sphere, it follows that each edge in G determines an edge of the 1-skeleton of the θ_1-eigenpolytope. Thus every distance regular graph is a spanning subgraph of the 1-skeleton of its θ_1-eigenpolytope. As a simple corollary we see that if θ_1 has multiplicity 3 then G is planar, because it is a spanning subgraph of the 1-skeleton of a 3-polytope.

Let G be distance regular with diameter d, let θ be a non-trivial eigenvalue of G and let u and v be two vertices at distance r in G. Let f_u be the function which maps a vertex at distance i from u to w_i, for $i = 0, \ldots, d$. We will call f_u the *standard* eigenvector for θ relative to the vertex u. It is not too hard to verify that f_u is an eigenvector for G with eigenvalue θ that takes its maximum value 1 on u. Let f_v be the standard eigenvector for θ relative to the vertex v, where v is at distance r from u in G, and consider the eigenvector $f_u + f_v$. Its value on u and on v is $1 + w_r$, so uv is an edge in the θ-eigenpolytope if $1 + w_r > w_i + w_j$ whenever $1 \le i, j \le d$. If θ_1 is the second largest eigenvalue of G, then $2w_1 \ge w_i + w_j$ whenever i and j are not zero. In particular, we see that if u and v are at distance 2 in G and

$$1 + w_2 > 2w_1, \qquad (9.1)$$

then uv is an edge in the θ_1-eigenpolytope of G. From Theorem 5.4, if G contains an induced 4-cycle, then $1 - 2w_1 + w_2 \ge 0$. This indicates that if G contains a 4-cycle, then it is likely not to coincide with the 1-skeleton of its θ_1 eigenpolytope.

There is a second simple reason why a distance regular graph might fail to be the 1-skeleton of its θ_1 eigenpolytope. By Theorem 7.1, the 1-skeleton of an m-polytope is m-connected. So if the valency k of G is less than the multiplicity m of θ_1, then G cannot possibly be a 1-skeleton. This condition is quite powerful, because if $k > m$ then Theorem 6.2 and (6.2) determine the least eigenvalue of the graph induced by the neighbourhood of a vertex in G, and if $k = m$ then the eigenpolytope must be simple. If $k < m$, there is no hope.

The detailed story is supplied by our next result, from [9]. If G is the Hamming graph $H(n, 2)$ then G_2 has exactly two connected components, which are isomorphic, and are known as *halved n-cubes*.

Theorem 9.1. *Let G be distance regular and let $\mathcal{P}$ be the eigenpolytope associated to the second-largest eigenvalue of G. Then G is the 1-skeleton of $\mathcal{P}$ if and only if it is one of the following:*
(a) a Johnson graph $J(v, \ell)$,
(b) a Hamming graph $H(n, q)$,
(c) a halved n-cube,
(d) a cycle,
(e) the Schläfli graph, the icosahedron, or the dodecahedron.

We make some comments on the proof. (For details, see [8].) One task is to verify that the graphs listed are indeed isomorphic to their 1-skeletons; this is comparatively straightforward. The more difficult part is to show that these are the only possibilities.

Suppose that G is distance regular with valency k and that θ_1 has multiplicity m. Recall that c_2 denotes the number of common neighbours of two vertices at distance 2 in G. Let $\mathcal{P}$ be the θ_1-eigenpolytope of G. As we noted above, the 1-skeleton of $\mathcal{P}$ is m-connected, whence $k \geq m$. By the results from Brouwer et al. [3: Chapter 3], it is possible to reduce to the case where $c_2 = 1$ and $k = m$.

If $k = m$, then $\mathcal{P}$ is simple (by Theorem 7.2) and every path of length 2 lies in a unique 2-face (by Theorem 7.3). As $c_2 = 1$, no 2-face is a 4-gon. If every 2-face is a triangle, then G is complete, and is hidden in the statement of the theorem as $J(v, 1)$. Thus we may assume that there are vertices x, y and z such that (x, y, z) is an induced path of length 2, and that this path is contained in a 2-face that is an n-gon. Therefore the angle between the vectors $\rho(x) - \rho(y)$ and $\rho(y) - \rho(z)$ is

$$\pi - \frac{2\pi}{n}.$$

On the other hand, using the identities from Section 5, we find that the cosine of the angle between this pair of vectors is

$$\frac{1 - 2w_1 + w_2}{2(1 - w_1)} = \frac{b_1 - 1 - \theta_1}{2b_1}$$

and therefore

$$\cos\left(\frac{2\pi}{n}\right) = \frac{\theta_1 + 1 - b_1}{2b_1}. \tag{9.2}$$

This shows that each 2-face of $\mathcal{P}$ is either a triangle or an n-gon, where n is determined by b_1 and θ.

Now consider the case when there are no triangles. Then $b_1 = k - 1$ and, as $\theta < k$, it follows from (9.2) that $\cos(2\pi/n) < 1/(k-1)$. Therefore if $k \geq 5$ then $n \geq 6$. But any 3-face of $\mathcal{P}$ is a simple 3-polytope, and therefore the 1-skeleton of any 3-face is a cubic planar graph. By Euler's formula, such a graph must have a face of size at most 5. Hence $k \leq 4$. If $k = 2$, then G is a cycle. If $k = 3$, then $n = 5$ and so G is a simple 3-polytope in $\mathbf{R}^3$ with all faces 5-gons; thus it is the dodecahedron. If $k = 4$, then $n = 5$ and Euler's formula implies that each face of $\mathcal{P}$ is a dodecahedron and hence $\mathcal{P}$ is a regular 4-polytope. The only regular 4-polytope with no triangle in its 1-skeleton is the 4-cube, $H(4, 2)$.

Similar geometric arguments can be used to treat the case when $a_1 > 0$. Full details appear in [8].

10. Questions

Perhaps despite appearances, the work in the last section was motivated by an attempt to determine the faces of the θ_2-eigenpolytope of $J(v, \ell)$; a failed attempt. We consider why this might be interesting.

If C is a subset of the vertices of a graph G, let C_i denote the set of vertices in G at distance i from C. The maximum integer r such that $C_r \neq \emptyset$ is the *covering radius* of C. The partition of $V(G)$ with cells C_i is the *distance partition* of G relative to C. We call C *completely regular* if its distance partition is equitable. Any vertex in a distance regular graph forms a completely regular subset, and it is not too hard to see that if C is completely regular with covering radius r, then C_r is also completely regular.

Suppose that G is distance regular and that C is a completely regular subset of G. It can be shown that if ρ is a representation of G belonging to the eigenvalue θ, then either $\rho(C)$ is a spherical 1-design, or it forms a face in the θ-eigenpolytope of G. The least non-negative integer t such that $\rho(C)$ is a face of the θ_{t+1}-eigenpolytope is the *strength* of C. If G is $J(v, \ell)$ and C is a subset of its vertices with strength t, then C is the set of blocks of a t-(v, ℓ, λ) design, for some λ. In the Hamming graphs we get orthogonal arrays.

The faces of the θ_1-eigenpolytope of $J(v, \ell)$ are implicitly determined by Meyerowitz in [16], and explicitly given in [8]. More on completely regular subsets will be found in Chapters 11 of [6] and [3], and in [13, 14, 18]. It is noted in [8] that any regular graph on v vertices determines a face in the θ_2-eigenpolytope of $J(v, 2)$, while Steiner triple systems on v points provide faces in the θ_3-eigenpolytope of $J(v, 3)$. This indicates some difficulties.

The evidence at hand suggests that the 1-skeleton of an eigenpolytope of a distance regular graph is often a complete graph. This raises two problems. The first is to find more classes of examples where the 1-skeleton is, if not isomorphic to G, at least not complete. The Chang graphs provide examples of the latter, as noted in [8].

On the other hand, we prove in [8] that some eigenpolytopes are even 3-neighbourly—that is any triple of vertices forms a face; for information about neighbourly polytopes, see [2: §14]. It is known that there are m-polytopes that are $\lfloor m/2 \rfloor$-neighbourly, but it is not clear that this can happen with eigenpolytopes of distance regular graphs.

References

[1] M. Balinski, On the graph structure of convex polyhedra in n-space, *Pacific J. Math.*, **11** (1961), 431–434.

[2] A. Brøndsted, *An Introduction to Convex Polytopes*. Springer (New York) 1983.

[3] A. E. Brouwer, A. M. Cohen and A. Neumaier, *Distance-regular Graphs*. (Springer, Berlin) 1989.

[4] P. Delsarte, J.-M. Goethals and J. J. Seidel, Spherical codes and designs, *Geom. Dedicata*, **6** (1977), 363–388.

[5] C. D. Godsil, Graphs, groups and polytopes, in: *Combinatorial Mathematics*, (edited by D. A. Holton and Jennifer Seberry), Lecture Notes in Mathematics 686, Springer, Berlin 1978, pp. 157–164.

[6] C. D. Godsil, Bounding the diameter of distance regular graphs, *Combinatorica*, **8** (1988), 333–343.

[7] C. D. Godsil, *Algebraic Combinatorics*. (Chapman and Hall, New York) 1993.

[8] C. D. Godsil, Equitable partitions, in *Combinatorics, Paul Erdős is Eighty, Vol. I*, (edited by D. Miklós, V. T. Sós, T. Szőnyi). (János Bolyai Mathematical Society, Budapest) 1993, pp. 173–192.

[9] C. D. Godsil, Eigenpolytopes of distance regular graphs, University of Waterloo Research Report CORR 94-12, 1994.

[10] C. D. Godsil and J. H. Koolen, On the multiplicity of eigenvalues of distance-regular graphs, *Linear Algebra Appl.*, to appear.

[11] C. D. Godsil and W. J. Martin, Quotients of association schemes, *J. Combinatorial Theory, Series A*, to appear.

[12] J. Koolen, *Euclidean Representations and Substructures of Distance-Regular Graphs*. Ph. D. Thesis, Technical University Eindhoven, 1994.

[13] C. Licata and D. L. Powers, A surprising property of some regular polytopes, *Scientia*, **1** (1988), 73–80.

[14] W. J. Martin, *Completely Regular Subsets*. Ph. D. Thesis, University of Waterloo, 1992.

[15] W. J. Martin, Completely regular designs of strength one, *J. Algebraic Combinatorics*, **3** (1994), 177–185.

[16] W. J. Martin and R. R. Zhu, On the classification of distance-regular graphs by eigenvalue multiplicities, submitted, (1993).

[17] A. Meyerowitz, Cycle-balanced partitions in distance-regular graphs, submitted.

[18] A. Neumaier, Characterisation of a class of distance-regular graphs, *J. reine angew. Math.*, **357** (1985), 182–192.

[19] A. Neumaier, Completely regular codes, *Discrete Math.*, **106/107** (1992), 352–360.

[20] D. L. Powers, The Petersen polytopes, Technical Report, Clarkson University, 1986.

[21] D. L. Powers, Eigenvectors of distance-regular graphs, *SIAM J. Matrix Anal. Appl.*, **9** (1988), 399–407.

[22] P. Terwilliger, Eigenvalue multiplicities of highly symmetric graphs, *Discrete Math.*, **41** (1982), 295–302.

[23] P. Terwilliger, The Johnson graph $J(d,r)$ is unique if $(d,r) \neq (2,8)$, *Discrete Math.*, **58** (1986), 175–189.

[24] P. Terwilliger, A new feasibility condition for distance-regular graphs, *Discrete Math.*, **61** (1986), 311–315.

[25] P. Terwilliger, Root systems and the Johnson and Hamming graphs, *European J. Combinatorics* **8** (1987), 73–102.

[26] R. R. Zhu, *Distance-Regular Graphs and Eigenvalue Multiplicities.* Ph. D. Thesis, Simon Fraser University, 1989.

[27] R. R. Zhu, The distance-regular graphs with an eigenvalue of multiplicity four, *J. Combinatorial Theory, Series B*, **57** (1993), 157–182.

LARGE SETS OF STEINER TRIPLE SYSTEMS

T.S.GRIGGS, UNIVERSITY OF CENTRAL LANCASHIRE
A.ROSA, MCMASTER UNIVERSITY

Dedicated to the memory of the Rev. T. P. Kirkman (1806-1895).

1. INTRODUCTION

This year, 1995, is the centenary of the death of the Reverend Thomas Pennington Kirkman, Rector of Croft and Fellow of the Royal Society. In mathematical circles he is known, amongst other things, for his 1847 paper [K1] in which he showed that Steiner triple systems of order v exist for all $v \equiv 1$ or $3 \pmod 6$. Since then, hundreds of papers have been written on many different aspects of such systems. Nevertheless, there are still fundamental but challenging questions which are unsolved. In this paper we consider just one of these concerned with the intersections of families of Steiner triple systems. We survey known results, present some recent advances, and pose a number of problems which suggest possible directions for future research in this area.

We start with some basic definitions. Recall that a Steiner triple system of order v (briefly $STS(v)$) is a pair $(V, \mathcal{B})$ where V is a v-set, and $\mathcal{B}$ is a collection of 3-subsets of V called *triples* such that each 2-subset of V is contained in exactly one triple. A family $(V, \mathcal{B}_1), ..., (V, \mathcal{B}_q)$ of q Steiner triple systems of order v, all on the same set V, is a *large set* of $STS(v)$ if every 3-subset of V is contained in at least one STS of the collection. Two $STS(v)$, $(V, \mathcal{B}_1), (V, \mathcal{B}_2)$ are *disjoint* if $\mathcal{B}_1 \cap \mathcal{B}_2 = \emptyset$, and *almost disjoint* if $|\mathcal{B}_1 \cap \mathcal{B}_2| = 1$. Interest in families of disjoint STS also dates back to the last century: Cayley [C] determined that the maximum number of disjoint $STS(7)$ is two and Sylvester [S] found a large set of 7 mutually disjoint $STS(9)$. This latter result too had been anticipated by Kirkman [K2]!

In this century, the problem of the existence of large sets of disjoint STS(v) was considered by Teirlinck [T1], Denniston [D] and Rosa [R]. Such a set necessarily contains $v - 2$ systems. Denniston's approach consisted of considering a permutation P on V having a $(v - 2)$-cycle and two fixed points, calculating all orbits of 3-subsets of V under P, and then constructing an STS(v) using exactly one triple from each orbit. The culmination of all this work was the series of classic papers by Lu [L] in which it was shown that a large set of disjoint STS(v) exists for all admissible $v \neq 7$ with six possible exceptions $v = 141,283,501,789,1501$ and 2365. The spectrum was finally completed again by Teirlinck [T2] who constructed large sets for the missing six values.

Some 20 years ago, Lindner and Rosa [LR] considered large sets of mutually almost disjoint (MAD) STS(v). They established that for all $v \equiv 1$ or 3 (mod 6), $v \geq 13$, any large set of MAD STS(v) contains $v - 1, v$, or $v + 1$ systems (with one extra possibility, 15 systems for $v=13$). Moreover, they constructed, for each $v \equiv 1$ or 3 (mod 6), a large set of v MAD STS(v). Subsequently, Phelps (unpublished) and Webb [W] established the impossibility of a large set of $v - 1$ MAD STS(v) of any order v. Because the proof of this result is not readily available but nevertheless is quite short, we include it here. It also gives information concerning the structure of a large set of v MAD STS(v).

Thus let $(V, \mathcal{B}_1), ..., (V, \mathcal{B}_q)$ be a large set of q MAD STS(v), let $\mathcal{D}_i$, $i = 1, ..., q$ be the collections of triples which occur in precisely i of the $\mathcal{B}_j$, $j = 1, ..., q$, and let $|\mathcal{D}_i| = k_i$. For a large set it then follows that

(A)
$$\sum_{i=1}^{q} k_i = \frac{1}{6} v(v - 1)(v - 2)$$

(B)
$$\sum_{i=1}^{q} i k_i = \frac{1}{6} q v(v - 1)$$

(C)
$$\sum_{i=2}^{q} \frac{1}{2} i(i - 1) k_i = \frac{1}{2} q(q - 1)$$

In addition, any triple $\{a, b, c\}$ can occur in at most $q + 3 - v$ $\mathcal{B}_j$s since there are $v - 3$ further triples of the form $\{a, b, x\}$, $x \neq c$. So $k_i = 0$ for $i > q + 3 - v$. When $q = v - 1$, (A) and (B) give $k_1 = \frac{1}{6}v(v - 1)(v - 3)$ and $k_2 = \frac{1}{6}v(v - 1)$. But (C) gives $k_2 = \frac{1}{2}(v - 1)(v - 2)$ which implies $v = 1$ or 3.

When $q = v$, the equations above yield $k_1 = \frac{1}{6}v(v - 1)(v - 3)$, $k_2 = 0$, $k_3 = \frac{1}{6}v(v-1)$. Next consider any 2-subset $\{a, b\} \in V$. It is contained in $v-2$ distinct triples, all of which must occur in at least one $\mathcal{B}_j$, $j = 1, ..., v$. Hence $v - 3$ of these must appear in just one $\mathcal{B}_j$, and the other in three $\mathcal{B}_j$s. Thus the collection $\mathcal{D}_3$ itself is a set of triples of an $STS(v)$, and, moreover, it must be distributed across the v systems according to an $STS(v)$. The method to achieve this in [LR] was to use a Steiner quadruple system $SQS(v + 1)$ (i.e. a Steiner system $S(3, 4, v + 1)$) on the set $\{0, 1, 2, ..., v\}$: if $q(xy)$ is the set of all triples of the derived triple system through x of the $SQS(v+1)$ in which then y is replaced by x, then let $\mathcal{B}_j = q(j0)$, $j = 1, ..., v$. This leads to our first question.

Problem 1. Do there exist large sets of v MAD $STS(v)$ other than those obtained from Steiner quadruple systems in the above way?

Phelps' and Webb's result left only the question of the existence of a large set of $v+1$ MAD $STS(v)$ in doubt. In this paper (Section 2 below) we present a general method for constructing large sets of $v + 1$ MAD $STS(v)$, and construct what we believe are the first examples of such sets (for orders $v = 13$ and 15). Our direct method is similar in spirit to that of Denniston used to give direct constructions of large sets of disjoint STS [D], even though it is, by necessity, more complicated.

2. A DIRECT CONSTRUCTION FOR LARGE SETS OF $v + 1$ MAD $STS(v)$

In this section, we develop a general outline for a direct construction of large sets of $v + 1$ MAD $STS(v)$. Let $\mathcal{S} = \{(V, \mathcal{B}_0), (V, \mathcal{B}_1), (V, \mathcal{B}_2), ..., (V, \mathcal{B}_v)\}$ be the large set of $v + 1$ MAD $STS(v)$ to be constructed. We identify V with Z_v, the additive group of residues modulo v. The group Z_v in its action on the set $\binom{V}{3}$ of all $\binom{v}{3}$ triples of V partitions $\binom{V}{3}$ into orbits

$O_1, ..., O_n$. Here n, the number of such orbits, equals $\lceil \frac{1}{6}(v-1)(v-2) \rceil$. When $v \equiv 1 \pmod 6$, all orbits are of full length v, but when $v \equiv 3 \pmod 6$, there is one short orbit of length $v/3$.

For a triple $T = \{a, b, c\}$ and $d \in Z_v$, we write $T + d = \{a+d, b+d, c+d\}$ mod v. If T_1, T_2 belong to the same orbit O_i, the *circular distance* $cd(T_1, T_2)$ between T_1 and T_2 equals $min(d_1, d_2)$ where $T_1 + d_1 = T_2$, $T_2 + d_2 = T_1$ (thus $cd(T_1, T_2) \leq \frac{1}{2}(v-1)$). If T_1, T_2 belong to different orbits, $cd(T_1, T_2)$ is undefined.

In the set $\mathcal{S}$, the STS $(V, \mathcal{B}_0)$ plays a special rôle. It is *cyclic* under Z_v, i.e. has a cyclic automorphism $\alpha = (0 \; 1 ... v-1)$, and as such comprises $\frac{1}{6}(v-1)$ (full) orbits of triples if $v \equiv 1 \pmod 6$, and $\frac{1}{6}(v-3)$ full orbits and the short orbit if $v \equiv 3 \pmod 6$. Let these full orbits be $Q_1, ..., Q_t$. The remaining v systems $(V, \mathcal{B}_1), ..., (V, \mathcal{B}_v)$ are pairwise isomorphic, and can be obtained from one another by applying an appropriate power of the cyclic automorphism α. Thus it suffices to describe the system $(V, \mathcal{B}_1)$ (say).

The STS(v) $(V, \mathcal{B}_1)$ consists, just as any STS(v), of $\frac{1}{6}v(v-1)$ triples. Since we want $|\mathcal{B}_0 \cap \mathcal{B}_1| = 1$, it follows that among the triples of $\mathcal{B}_1$ there should be *exactly one* representative of *exactly one* of the orbits $Q_1, ..., Q_t$, and no representative of the remaining $t-1$ orbits (or of the short orbit when $v \equiv 3 \pmod 6$). This will ensure that $|\mathcal{B}_0 \cap \mathcal{B}_i| = 1$ for all $i = 1, 2, ..., v$.

Assume now that $v \equiv 1 \pmod 6$ [when $v \equiv 3 \pmod 6$, the considerations are quite similar]. Then $t = \frac{1}{6}(v-1)$. Since no representative of $t-1$ orbits (from among $Q_1, ..., Q_t$) can occur in $\mathcal{B}_1$, this leaves

$$\frac{1}{6}(v-1)(v-2) - (\frac{1}{6}(v-1) - 1) = \frac{1}{6}(v-1)(v-3) + 1$$

orbits to be represented in $\mathcal{B}_1$. Each of these orbits must have at least one representative in $\mathcal{B}_1$, since $(V, \mathcal{B}_0), ..., (V, \mathcal{B}_v)$ is a large set of STS(v). But since $\frac{1}{6}v(v-1) - (\frac{1}{6}(v-1)(v-3) + 1) = \frac{1}{2}(v-3) > 0$, several of these orbits must have more than one representative in $\mathcal{B}_1$.

Whenever $\mathcal{B}_1$ contains two representatives of the same orbit whose circular distance is d, we have $|\mathcal{B}_i \cap \mathcal{B}_j| = 1$ whenever $i - j = \pm d$ mod v. Thus the set of circular distances between multiple representatives from the orbits of triples present in $\mathcal{B}_1$, together with their negatives, must comprise the set of non-zero residues modulo v.

The above considerations lead to the following theorem.

Theorem 2.1. *Let $(V, \mathcal{B}_0)$ be a cyclic $STS(v)$ where $\mathcal{B}_0$ consists of orbits $Q_1, ..., Q_s$ where $s = t$ or $t + 1$, and let $(V, \mathcal{B}_1)$ be an $STS(v)$ with the following properties:*

(i) *$\mathcal{B}_0 \cap \mathcal{B}_1 = T = \{a, b, c\}$ where T belongs to a full orbit of triples under Z_v;*

(ii) *there exists exactly one orbit O_0 of triples such that $\mathcal{B}_1$ contains exactly 3 representatives T_{01}, T_{02}, T_{03} of O_0;*

(iii) *there are exactly $u = \frac{1}{2}(v - 7)$ orbits of triples $O_1, ..., O_u$ such that $\mathcal{B}_1$ contains exactly two representatives T_{i1}, T_{i2}, $i = 1, ..., u$, of each of these orbits;*

(iv) *$\{cd(T_{01}, T_{02}), cd(T_{01}, T_{03}), cd(T_{02}, T_{03})\} \cap \{cd(T_{i1}, T_{i2}) : i = 1, 2, ..., u\} = \{1, 2, ..., \frac{1}{2}(v - 1)\}$.*

Then there exists a large set of $v + 1$ MAD $STS(v)$.

Proof. In view of the remarks preceding the statement of the theorem, it suffices to show that each orbit of triples of $\binom{V}{3}$ is represented in $\mathcal{B}_0 \cup \mathcal{B}_1$, and thus that $\{(V, \mathcal{B}_0), (V, \mathcal{B}_1), ..., (V, \mathcal{B}_v)\}$ is a large set of $STS(v)$. Conditions (i),(ii),(iii) imply that there are exactly $\frac{1}{6}v(v - 1) - v + 3$ triples in $\mathcal{B}_1$ other than T, T_{01}, T_{02}, T_{03}, T_{i1}, T_{i2}, $i = 1, 2, ..., u$, which is precisely the number of (full) orbits of triples of $\binom{V}{3}$ other than $Q_1, ..., Q_s, O_0, O_1, ..., O_u$; each of these remaining orbits will thus have exactly one representative in $\mathcal{B}_1$.□

We have implemented a search for an $STS(v)$ satisfying the conditions of Theorem 2.1 for $v = 13$ and $v = 15$ on a 486DX/33 computer. For $v = 13$, the search was complete; there is, up to an isomorphism, a unique solution of the type described in Theorem 2.1. This solution is displayed in Table 1. We list the triples of $\mathcal{B}_0$ and $\mathcal{B}_1$ explicitly, and indicate the orbit to which a particular triple of $\mathcal{B}_1$ belongs by listing the difference triple associated with the orbit. Somewhat curiously, the STS $(V, \mathcal{B}_1)$ is "*the other*" i.e. the non-cyclic STS(13)!

For $v = 15$, there appear to be a large number of solutions of this type. After only a tiny fraction of a complete search, 43 distinct solutions (i.e. large sets) were found, each taking PG(3,2) as the cyclic system $(V, \mathcal{B}_0)$. Of the 43 solutions, as many as 23 out of the 80 nonisomorphic STS(15) appeared as the system $(V, \mathcal{B}_1)$. Thereafter the search was discontinued. We estimate that the number of distinct solutions of this type for $v = 15$ is already in thousands. One solution, with the STS(15) $(V, \mathcal{B}_1)$ isomorphic to STS No.39 (cf. [MPR]) is given in Table 2.

Table 1

A large set of 14 MAD STS(13)

	Triples			Difference triples	
$\mathcal{B}_0$:	$\{0,1,4\}$,	$\{0,2,7\}$	mod 13	$\langle 1,3,4 \rangle$,	$\langle 2,5,6 \rangle$
$\mathcal{B}_1$:	$\{0,2,8\}$,	$\{2,7,9\}$	$\{4,8,10\}$	$\langle 5,2,6 \rangle$	
	$\{4,5,7\}$,	$\{8,9,11\}$		$\langle 1,2,3 \rangle$	
	$\{5,6,10\}$,	$\{7,8,12\}$		$\langle 1,4,5 \rangle$	
	$\{1,5,8\}$,	$\{2,5,11\}$		$\langle 4,3,6 \rangle$	
	$\{2,3,4\}$			$\langle 1,1,2 \rangle$	
	$\{0,1,4\}$			$\langle 1,3,4 \rangle$	
	$\{0,6,7\}$			$\langle 1,4,5 \rangle$	
	$\{4,11,12\}$			$\langle 1,5,6 \rangle$	
	$\{4,9,10\}$			$\langle 5,1,6 \rangle$	
	$\{1,2,10\}$			$\langle 4,1,5 \rangle$	
	$\{7,10,11\}$			$\langle 3,1,4 \rangle$	
	$\{0,10,12\}$			$\langle 2,1,3 \rangle$	
	$\{4,6,8\}$			$\langle 2,2,4 \rangle$	
	$\{0,3,11\}$			$\langle 2,3,5 \rangle$	
	$\{1,3,7\}$			$\langle 2,4,6 \rangle$	
	$\{3,5,12\}$			$\langle 4,2,6 \rangle$	
	$\{1,9,12\}$			$\langle 3,2,5 \rangle$	
	$\{3,6,9\}$			$\langle 3,3,6 \rangle$	
	$\{2,6,12\}$			$\langle 3,4,6 \rangle$	
	$\{1,6,11\}$			$\langle 3,5,5 \rangle$	
	$\{0,5,9\}$			$\langle 4,4,5 \rangle$	

Table 2

A large set of 16 MAD STS(15)

Triples			Difference triples
$\mathcal{B}_0$: {0,1,4}	mod 15		⟨1,3,4⟩
{0,2,8}	mod 15		⟨2,6,7⟩
{0,5,10}	mod 15		⟨5,5,5⟩
$\mathcal{B}_1$: {0,6,8}	{1,7,9},	{3,9,11}	⟨6,2,7⟩
{8,9,13}	{0,1,5}		⟨1,4,5⟩
{6,10,11}	{0,10,14}		⟨4,1,5⟩
{4,7,10}	{0,9,12}		⟨3,3,6⟩
{1,6,14}	{5,7,12}		⟨2,5,7⟩
{4,5,8}			⟨1,3,4⟩
{11,12,13}			⟨1,1,2⟩
{1,2,4}			⟨1,2,3⟩
{4,13,14}			⟨1,5,6⟩
{6,7,13}			⟨1,6,7⟩
{2,9,10}			⟨1,7,7⟩
{3,4,12}			⟨6,1,7⟩
{2,7,8}			⟨5,1,6⟩
{2,3,14}			⟨3,1,4⟩
{3,5,6}			⟨2,1,3⟩
{0,2,13}			⟨2,2,4⟩
{4,6,9}			⟨2,3,5⟩
{1,10,12}			⟨2,4,6⟩
{3,8,10}			⟨5,2,7⟩
{8,12,14}			⟨4,2,6⟩
{1,3,13}			⟨3,2,5⟩
{0,3,7}			⟨3,4,7⟩
{1,8,11}			⟨3,5,7⟩
{2,5,11}			⟨3,6,6⟩
{5,10,13}			⟨5,3,7⟩
{7,11,14}			⟨4,3,7⟩
{0,4,11}			⟨4,4,7⟩
{5,9,14}			⟨4,5,6⟩
{2,6,12}			⟨5,4,6⟩

Based on the above results, we expect sets of $v+1$ MAD STS(v) to exist for all $v \equiv 1$ or 3 (mod 6), $v \geq 13$. Unfortunately, we know of no further direct constructions other than the one outlined in this section, nor of any recursive construction which will likely be needed to prove this. We can present this as our second question.

Problem 2. Do there exist large sets of $v+1$ MAD STS(v) for all $v \equiv 1$ or 3 (mod 6), $v \geq 13$?

3. LARGE SETS OF $v - 1$ NEARLY DISJOINT STS(v)

The minimum number of STS(v) in a large set is $v - 2$ in which case they are disjoint. Phelps' and Webb's result ensures that if the systems are mutually almost disjoint (MAD), their minimum number is v. This naturally raises the question of whether it is possible to obtain large sets of $v - 1$ STS(v) if we allow every distinct pair of them to be either disjoint or almost disjoint. This motivates the following definition. A family $(V, \mathcal{B}_0), ..., (V, \mathcal{B}_q)$ of STS(v) is said to be *nearly disjoint* if $|\mathcal{B}_i \cap \mathcal{B}_j| \leq 1$ for $i, j = 1, ..., q, i \neq j$. In this section we develop a recursive construction to show that there exists a large set of $v - 1$ nearly disjoint (ND) STS(v) for infinitely many values of v. We begin with an example.

Example 1. Let $V = \{1, 2, 3, 4, 5, 6, 7\}$, and consider the following 6 sets of triples of STS(7) on V:

$\mathcal{B}_1$	$\mathcal{B}_2$	$\mathcal{C}_1$	$\mathcal{C}_2$	$\mathcal{C}_3$	$\mathcal{C}_4$
123	123	124	125	126	127
145	146	137	136	135	134
167	157	156	147	147	156
246	247	235	234	237	236
257	256	267	267	245	245
347	345	346	357	346	357
356	367	457	456	567	467

In the above example, $|\mathcal{B}_1 \cap \mathcal{B}_2| = 1$, $|\mathcal{C}_i \cap \mathcal{C}_j| = 1$ for $i, j \in \{1, 2, 3, 4\}$, $i \neq j$, $|\mathcal{B}_i \cap \mathcal{C}_j| = 0$ for $i \in \{1, 2\}$, $j \in \{1, 2, 3, 4\}$. Moreover, the union of the sets of triples of the 6 STS above contains all 3-subsets of $\{1, 2, 3, 4, 5, 6, 7\}$. Thus the family of the 6 STS above is a large set of 6 nearly disjoint STS(7). It is easily seen that this large set is unique up to an isomorphism.

For which other values does a large set of $v - 1$ nearly disjoint STS(v) exist? If such a set is to exist for $v = 9$ then from among the 28 pairs of systems that may be obtained from the 8 STS(9) in such a set, there must be 16 pairs that are mutually disjoint, and 12 pairs that are almost disjoint.

However, here we are out of luck: an exhaustive computer search reveals that there exists no such large set.

However, we are able to present a recursive construction which shows that there are infinitely many values of v for which a large set of $v-1$ nearly disjoint STS(v) does exist.

Assume that there exists a large set of nearly disjoint STS(v) $\{(V, \mathcal{B}_1), ..., (V, \mathcal{B}_{v-1})\}$ where $V = \{a_1, ..., a_v\}$. Let $W = V \cup Z_{v+1}$; we are going to construct a large set of $(v-1) + (v+1) = 2v$ STS($2v + 1$) on W: $\{(W, \mathcal{B}_1^*), ..., (W, \mathcal{B}_{v-1}^*), (W, \mathcal{C}_0), (W, \mathcal{C}_1), ..., (W, \mathcal{C}_v)\}$.

(1) To construct the $\mathcal{C}_i$s, take first the near-1-factorization $\mathcal{F}_0$ of K_v on $V = \{a_1, ..., a_v\}$ given by $\mathcal{F}_0 = \{f_{01}, f_{02}, ..., f_{0v}\}$ where $f_{0i} = \{\{a_i a_{i+1}, a_{i+2} a_{i+v-1}, a_{i+3} a_{i+v-2}, ..., a_{i+(v-1)/2}, a_{i+(v+3)/2}\} : i = 1, 2, ..., v\}$ (subscripts reduced modulo v to the range $\{1, 2, ..., v\}$; the isolated vertex of f_{0i} is $a_{i+(v+1)/2}$). For $i \neq j$, $i = 1, 2, ..., v$; $j \in Z_{v+1}$, let $f_{ji} = f_{0,i-j}$ (subscripts reduced modulo $v + 1$ to the range $\{0, 1, ..., v\}$), and put $\mathcal{F}_j = \{f_{j1}, f_{j2}, ..., f_{jv}\}$ for $j \in Z_{v+1}$. Put $\mathcal{D}_0 = \{\{a_p, a_q, i\} : \{a_p, a_q\} \in f_{0i}\}$ (we have $|\mathcal{D}_0| = \frac{1}{2}v(v-1)$), and $\mathcal{D}_j = \{\{a_p, a_q, i\} : \{a_p, a_q\} \in f_{ji}\}$.

It is easily seen that an element $a_i \in V$ does not occur in a triple of $\mathcal{D}_0$ with exactly two elements of Z_{v+1}, namely 0 and $i + \frac{1}{2}(v-1)$. Therefore let $\mathcal{E}_0 = \{\{a_i, 0, i + (v-1)/2\} : i = 1, 2, ..., v\}$, and for $j \in Z_{v+1}$, let $\mathcal{E}_j = \{\{a_i, j, i + j + (v-1)/2\} : i = 1, 2, ...v\} \bmod (v+1)$.

Take now an SQS($v + 1$) on Z_{v+1}, and use it to construct an overlarge set of $v + 1$ disjoint STS(v) (cf [LR]; for a definition of an overlarge set, see, e.g. [SS]), say, $\{(Z_{v+1} \backslash \{i\}, \mathcal{G}_i), i \in Z_{v+1}\}$. Let now $\mathcal{C}_i = \mathcal{D}_i \cup \mathcal{E}_i \cup \mathcal{G}_i$. Then $\{(W, \mathcal{C}_i) : i \in Z_{v+1}\}$ is a set of $v + 1$ nearly disjoint STS($2v + 1$). Indeed, each triple of $\mathcal{D}_i$ contains two elements of V and one element of Z_{v+1}; for fixed p, q, each set $\mathcal{D}_j$ contains exactly one triple of the form $\{a_p, a_q, i\}$, and for each such i, there is a unique set among the $\mathcal{D}_j$s that contains $\{a_p, a_q, i\}$. Thus $\mathcal{D}_i \cap \mathcal{D}_j = \emptyset$ for $i \neq j$. We also have $\mathcal{G}_i \cap \mathcal{G}_j = \emptyset$ by definition of an overlarge set. The only intersections occur among the $\mathcal{E}_i$s. In fact, the specific near-1-factorizations above guarantee that $|\mathcal{E}_i \cap \mathcal{E}_j| = 1$ if $i - j \equiv \frac{1}{2}(v+1)$, and $\mathcal{E}_i \cap \mathcal{E}_j = \emptyset$ otherwise.

(2) Assume now $v = 2^n - 1$ for some positive integer n. Let $\mathcal{H} = \{h_1, ..., h_v\}$ be a 1-factorization of K_{v+1} on Z_{v+1} obtained by partitioning the set of edges $E_d = \{\{x, y\} : x - y = \pm d\}$, for $d = 1, ..., \frac{1}{2}(v+1)$, into two 1-factors where h_i is the 1-factor containing the edge $\{0, i\}$ [if E_d has more than one component, there is more than one way to partition E_d into two 1-factors].

Consider now the cyclic latin square $L = (l_{ij})$ of order v:

1	2	3	...	$v-2$	$v-1$	v
v	1	2	...	$v-3$	$v-2$	$v-1$
$v-1$	v	1	...	$v-4$	$v-3$	$v-2$
....						
3	4	5	...	v	1	2
2	3	4	...	$v-l$	v	1

Delete the $\frac{1}{2}(v+1)$-st row (the "middle" row), and consider the remaining $v - 1$ rows. For simplicity, let these $v - 1$ rows be labelled by integers $1, 2, ..., v - 1$. For $i = 1, ..., v - 1$, form the set of triples $\mathcal{B}'_i = \{\{x, y, a_t\} : \{x, y\} \in h_j, l_{ij} = t\}$.

Let $\mathcal{B}^*_i = \mathcal{B}_i \cup \mathcal{B}'_i, i = 1, ..., v - 1$. Then $\{(W, \mathcal{B}^*_i), i = 1, ..., v - 1\}$ is a set of $v - 1$ nearly disjoint STS$(2v + 1)$. Indeed, since L is a latin square, $\mathcal{B}'_p \cap \mathcal{B}'_q = \emptyset$ for $p \neq q$, and by our assumption, $|\mathcal{B}_p \cap \mathcal{B}_q| \leq 1$ for $p \neq q$.

(3) The set $\{(W, \mathcal{B}^*_1), ..., (W, \mathcal{B}^*_{v-1}), (W, \mathcal{C}_0), ..., (W, \mathcal{C}_v)\}$ is clearly a large set of STS, since every 3-subset of W is a triple in at least one system. Thus in order to show that it is a large set of $2v$ nearly disjoint STS$(2v + 1)$, it only remains to be shown that $|\mathcal{B}_i \cap \mathcal{C}_j| \leq 1$ for any $i \in \{1, ..., v - 1\}$ and any $j \in Z_{v+1}$. This is straightforward to verify, due to our choice of the 1-factorization $\mathcal{H}$.

Let us remark that the set of $v+1$ nearly disjoint STS$(2v + 1)$ given in part (1) of the above construction does not depend on the fact that $v = 2^n - 1$ for some n, and so it could, in principle, be a part of a desired (and desirable) general $v \rightarrow 2v+1$ construction. But at the time of writing, we did not succeed, in the general case, to construct, and to tie up with the set constructed in (1), a 1-factorization of the kind of the 1-factorization $\mathcal{H}$ given in part (2) in such a way that the requirement in part (3) be satisfied.

Nevertheless, the construction above ensures the following.

Theorem 3.1. *A large set of $v-1$ nearly disjoint STS(v) exists for all v of the form $v = 2^n - 1$, where n is a positive integer.*

It also allows us to ask our third question.

Problem 3. For which orders $v \equiv 1$ or 3 (mod 6) do there exist large sets of $v - 1$ ND STS(v)?

4. RELATED QUESTIONS AND OPEN PROBLEMS

In any large set of $v - 2$ pairwise disjoint STS(v), each triple of the underlying set V occurs the same number of times, namely once. This automatically leads to the generalization of whether we can construct large sets in which each triple occurs the same number $x > 1$ times, apart, of course, from repeating a large set of disjoint STS x times which is clearly "cheating". One example we can give is the unique maximal large set of 15 MAD STS(7)s (cf. [LR]). In this, every triple occurs three times.
So we have our fourth question.

Problem 4. For which values of $x > 1$ and orders $v \equiv 1$ or 3 (mod 6) do there exist large sets of STS(v) in which each triple occurs exactly x times, and in which no other large set in which each triple occurs the same number $y < x$ times is contained?

Continuing further, if in any large set of STS(v) each triple of the underlying set occurs with one of two frequencies x and $y > x$ then it is easily seen that the collection of all triples occurring y (and x) times themselves form a λ-fold triple system for some λ. It was shown in Section 1 that any large set of v MAD STS(v) has the property that each triple of the underlying set occurs either once or three times, and the set of triples occuring three times forms an STS(v) (cf. [LR]). The large sets of $v + 1$ MAD STS(v) constructed in Section 2 no longer have this nice property: a triple may occur once, twice or three times in such a set. On the other hand, in the large sets of $v - 1$ nearly disjoint STS(v) constructed in Section 3, each triple occurs once or twice, and again it is easy to see that the triples occurring twice form an STS(v). The large set of 13 nearly disjoint STS(13) in Example 2 below also has this property. Each triple occurs once or twice, the latter forming a twofold triple system, TTS(13).

Example 2. A large set of 13 nearly disjoint STS(13)

$V = Z_{13}$,
$\mathcal{B}_0$: $\{0,2,8\}, \{1,3,8\}, \{3,9,11\}, \{3,5,6\}, \{5,7,8\}, \{4,8,9\}, \{1,2,10\},$
$\{2,6,12\}, \{2,5,9\}, \{0,3,12\}, \{2,3,4\}, \{0,10,11\}, \{0,1,5\}, \{4,5,10\},$
$\{5,11,12\}, \{1,6,7\}, \{6,9,10\}, \{8,10,12\}, \{6,8,11\}, \{0,7,9\}, \{0,4,6\},$
$\{1,9,12\}, \{1,4,11\}, \{4,7,12\}, \{3,7,10\}, \{2,7,11\};$
$\mathcal{B}_i = \mathcal{B}_0 + i \bmod 13, i = 1, ..., 12.$

If this example is any indication, large sets of STS in which each triple of the underlying set appears with frequency x or y are likely to exist in abundance. This gives our fifth problem.

Problem 5. For which values of x and y, $1 \le x < y$, and orders $v \equiv 1$ or 3 (mod 6) do there exist large sets of STS(v) in which each triple occurs with frequency either x or y?

We observe that the case $x = 1, y = 3$ is completely solved, and that the case $x = 1, y = 2$ for which there are partial results above, might be a sensible place to start.

One may ask similar questions concerning the number of triples in which two STS of a large set intersect. In particular, if $t \ge 2$, we know of no example of a *large* set of STS such that any two systems intersect in exactly t triples. We call such systems *uniformly intersecting* large sets. It seems at first glance that one should be able to use projective spaces over GF(2) to provide such examples but it does not seem to be the case.

We can make this our sixth question.

Problem 6. Do uniformly intersecting large sets of STS(v) exist for $t \ge 2$?

The large set in Example 2 is *minimal* (as are all large sets constructed in Sections 2 and 3) in that if any of the STS in the set is deleted, the set is no longer large. Again, if one allows the intersection size to vary, there is likely to be a multitude of minimal large sets even for small orders, as the following theorem seems to indicate.

Theorem 4.1. *Let S be a minimal large set of q STS(7)s. Then S is one of the following:*

(i) *q = 6; the set of 6 STSs consists of two subsets A and B, respectively, of 2 and 4 STSs. The 2 systems of A are MAD, as are the 4 systems of B, and any system of A and any system of B are disjoint (cf. Example 1).*

(ii) *q = 7; the set of 7 STSs consists of MAD STSs from an SQS(8) (cf.[LR]).*

(iii) *q = 8; the set of 8 consists of an STS, and any 7 (i.e. all but one) STSs disjoint from it; these 7 STSs are MAD.*

(iv) *q = 8; the set of 8 consists of an STS, six STSs disjoint from it, and one STS intersecting it in 3 triples; the latter 7 STSs are MAD.*

(v) *q = 10; the set of 10 is MAD, and is obtained as follows: delete one STS from a minimal large set of 7 MAD STSs (see (ii)), and add four STSs from the corresponding (i.e. the same) 15-clique of MAD STSs so as to cover the 4 "missing" triples.*

In each of the above cases (i)-(v), the minimal large set is unique up to an isomorphism.

Another interesting concept is that of *maximal* sets of disjoint (almost disjoint, nearly disjoint) STSs. Of course, there is only one maximal set of (two) disjoint STS(7). It has been known for quite some time that there are exactly two nonisomorphic large sets of 7 disjoint STS(9), with automorphism groups of order 42 and 54, respectively (cf. [B], [KM]). Maximal sets of disjoint STS(9) have recently been classified by Cooper [C1]:

Any maximal set of q disjoint STS(9) has $q = 4$, 5 or 7.

(i) There is up to an isomorphism a unique maximal set of 4 STS(9):

123	127	128	129
847	635	534	835
965	489	976	764

The automorphism group of this set has order 1.

(ii) There are exactly 3 nonisomorphic maximal sets of 5 STS(9):

No.1	123	124	125	126	128
	847	938	937	538	935
	965	765	684	947	476
No.2	123	124	125	126	127
	847	938	937	438	435
	965	765	684	957	698
No.3	123	124	125	127	128
	847	938	634	534	635
	965	765	987	896	479

The automorphism groups of the three sets above have order 2, 2, and 20, respectively. [The 12 triples of each of the above STS(9) are given by the three rows, three columns, and six diagonals.]

Cooper [C1] has also produced a multitude of maximal sets of disjoint STS(13) of size 7,8 and 9.

As for the general case, the results of [CH] can be reinterpreted in terms of maximal sets of disjoint STSs:

There exists a maximal set of $v-4$ disjoint STS(v) whenever $v = 5 \cdot 2^a - 1$, $a \geq 1$, and a maximal set of $v - 5$ disjoint STS(v) whenever $v = 5 \cdot 2^a - 1$, $a \geq 1$, or $v = 2^b - 1$, $b \geq 3$.

For $v > 7$, a conjectured lower bound of $\frac{1}{2}(v - 1)$ for the minimum number of STSs in a maximal set of disjoint STS(v) would agree with the conjectured upper bound $\frac{1}{2}(v - 3)$ for the index λ of an indecomposable triple system of order v. However, this has been proved only for a single order, namely $v = 9$!

It is also known (cf. [LR]) that a maximal set of MAD STS(9) can be of size 4,5 and 9, and that any set of v MAD STS(v) is maximal. To the best of our knowledge, that is the extent of what is known here.

There is much work to be done.

References

[B]   S.Bays, *Une question de Cayley relative au problème des triades de Steiner*, Enseignement Math. (1917), no. 19, 57–67.

[C]   A.Cayley, *On the triadic arrangements of seven and fifteen things*, London, Edinburgh and Dublin Philos. Mag. and J. Sci. **3** (1850,), no. 37, 50–53.

[C1]   D.S.Cooper, *Maximal disjoint Steiner triple systems of orders nine and thirteen*, (to appear).

[CH]   C.J.Colbourn, J.J.Harms, *Partitions into indecomposable triple systems*, Ann. Discrete Math. (1987), no. 34, 107–118.

[D]   R.H.F.Denniston, *Some packings with Steiner triple systems*, Discrete Math. (1974), no. 9, 213–227.

[K1]   T.P.Kirkman, *On a problem in combinations*, Cambridge and Dublin Math. J. (1847), no. 2, 141–204.

[K2]   T.P.Kirkman, *Note on an unanswered prize question*, Cambridge and Dublin Math. J. (1850), no. 5, 255–262.

[KM]   E.S.Kramer, D.M.Mesner, *Intersections among Steiner systems*, J. Combinat. Theory (A) (1974), no. 16, 273–285.

[LR]   C.C.Lindner, A.Rosa, *Construction of large sets of almost disjoint Steiner triple systems*, Canad. J. Math. (1975), no. 27, 256–260.

[MPR] R.A.Mathon, K.T.Phelps, A.Rosa, *Steiner triple systems and their properties*, Ars Combinat. (1983), no. 15, 3–100.

[L]   J.X.Lu, *On large sets of disjoint Steiner triple systems I-III*, J. Combinat. Theory (A) (1983), no. 34, 140–182.
       On large sets of disjoint Steiner triple systems IV-VI, J. Combinat. Theory (A) (1984), no. 37, 136–192.

[R]   A.Rosa, *A theorem on the maximum number of disjoint Steiner triple systems*, J. Combinat. Theory (A) (1975), no. 18, 305–312.

[S]   J.J.Sylvester, *Remark on the tactic of nine elements*, Phil. Mag. **4** (1861), no. 22, 144–147.

[SS]   M.J.Sharry, A.P.Street, *A doubling construction for overlarge sets of Steiner triple systems*, Ars Combinat. (1991), no. 32, 143–151.

[T1]   L.Teirlinck, *On the maximum number of disjoint Steiner triple systems*, Discrete Math. (1973), no. 6, 299–300.

[T2]   L.Teirlinck, *A completion of Lu's determination of the spectrum for large sets of disjoint Steiner triple systems*, J. Combinat. Theory (A) (1991), no. 57, 302–305.

[W]   T.M.Webb, *Some Constructions of Sets of Mutually Almost Disjoint Steiner Triple Systems*, M.Sc.Thesis, Auburn University, 1977.

SEARCHING WITH LIES

RAY HILL

Department of Mathematics & Computer Science
University of Salford
Salford M5 4WT
England

1. INTRODUCTION

Consider a game of Twenty Questions in which someone thinks of a number between one and one million. A second person is allowed to ask questions to each of which the first person is supposed to answer only yes or no. Since one million is just less than 2^{20}, it is clear that a "halving" strategy (i.e. asking "Is the number in the first half million?", and so on) will determine the number within twenty questions. But now suppose that up to some given number e of the answers may be lies. How many questions does one now need to get the right answer?

This is Ulam's searching problem, posed by Stanislaw Ulam (1976) in his autobiography "The Adventures of a Mathematician".

The problem has recently been solved for all values of e (for the cases of both 2^{20} and 10^6 objects). We give the solution in Figure 1 and an outline of the proof in Section 3.

More generally, we may consider the problem of finding the smallest number $f(M,e)$ of yes-no questions sufficient to determine one of M objects if up to e of the answers may be lies. In Section 4, we survey the present state of knowledge regarding this function.

In Section 5, we consider a version of Ulam's problem without feedback, where all the questions must be asked in advance of receiving any answers. This is equivalent to a problem in the theory of error-correcting codes.

Finally, in Section 6, we consider several other variations of Ulam's game, such as that in which the proportion of lies told at any point must not exceed some given value, and that in which only comparison questions may be asked.

Figure 1   The solution of Ulam's problem for M = 2^{20} and M = 10^6

e	$f(2^{20},e)$	
0	20	(trivial)
1	25	Pelc (1987)
2	29	Czyzowicz, Mundici and Pelc (1988)
3	33	Negro and Sereno (1992)
4	37	Hill and Karim (1992)
5	40	
6	43	
7	46	
8	50	Hill, Karim and Berlekamp (to appear)
9	53	
⋮	⋮	
e	3e+26	
⋮	⋮	

The value of $f(10^6,e)$ is the same as that of $f(2^{20},e)$ for all e except e = 4, where we have $f(10^6,4) = 36$.

---

2.   PRELIMINARY RESULTS AND TERMINOLOGY

The general problem of finding $f(M,e)$ had been considered somewhat earlier than Ulam's posing of the problem, by Berlekamp (1968) in the context of block coding for the binary symmetric channel with feedback. Berlekamp's paper, which was based on his doctoral thesis of 1964, provides most of the ingredients for our solution of Ulam's problem.

Consider a game played by two players, the **Responder** who chooses an integer x (called the **target**) from the set {1,2,...,M}, and the **Questioner** who has to determine this number in n queries of the form "Is x in the set S?", where S is some subset of the set {1,2,...,M}.

An **e-lie game** is one in which the Responder is allowed to lie at most e times. If the number n of questions allowed is also fixed, we refer to the game as an **(n,M,e) game**. Let f(M,e) be the minimum number of questions needed to guarantee a win for the Questioner in the e-lie game.

The state of knowledge of the Questioner at any stage of the game is given by a collection of disjoint subsets X_0, X_1, ..., X_e of {1,2,...,M}, where X_i is the subset of possible targets under the condition that the Responder has told exactly i lies (i=0,1,...,e). If we let x_i denote the cardinality of the subset X_i, then we define $(x_0, x_1, ..., x_e)$ to be the **state** of the game at this stage.

Now consider the question "Is the target in S?". Let u_i be the cardinality of the set $X_i \cap S$. We write such a query as $[u_0, u_1, ..., u_e]$. The effect of the query is to split the current state into two **reduced states**. The reduced state is $(u_0, u_1 + x_0 - u_0, u_2 + x_1 - u_1, ..., u_e + x_{e-1} - u_{e-1})$ if the answer is yes and $(x_0 - u_0, x_1 - u_1 + u_0, x_2 - u_2 + u_1, ..., x_e - u_e + u_{e-1})$ if the answer is no.

In recent papers, Spencer (1992), Spencer and Winkler (1992), Aslam and Dhagat (1991), the liar game has been re-interpreted as a **chip game**. Imagine a board with positions marked (from left to right) 0,1,2,... . There is one chip for each possible target. The game starts with M chips on level 0. At each stage the Questioner selects some chips on the board. These chips correspond to the subset S of {1,2,..., M} which the Questioner wishes to ask about. The Responder then either moves all the selected chips one position to the right, or moves those chips not selected one position to the right. Chips to be moved from position e are removed from the board. The Questioner wins if at the end of the game there is precisely one chip left on the board.

Note that asking the question $[u_0, u_1, ..., u_e]$ is equivalent to asking the complementary question $[v_0, v_1, ..., v_e]$, where $u_i + v_i = x_i$. We may represent the transition from a current state to a new state via the diagram

$$(x_0, x_1, ..., x_e)$$
$$[u_0, u_1, ..., u_e]$$
$$[v_0, v_1, ..., v_e]$$

$(u_0, u_1 + v_0, ..., u_e + v_{e-1})$ $\qquad\qquad$ $(v_0, v_1 + u_0, ..., v_e + u_{e-1})$.

Note that the reduced yes-state is given simply by adding u_i's to v_j's diagonally according to

$$
\begin{array}{ccccc}
u_0 & u_1 & u_2 \cdots & u_{e-1} & u_e \\
/ & / & / & & / \\
v_0 & v_1 & v_2 \cdots & v_{e-1} & v_e
\end{array}
$$

while the no-state is given via the reverse diagonals.

If, in a given game, only a certain number of questions are allowed, then the state at any stage is called an **n-state** if there are n questions left. A state with $x_i = 1$ for some i and $x_j = 0$ otherwise is called a **singlet**. A state with $x_i = 0$ for all i is called a **null state**. If for a given n-state, there exists a questioning strategy such that after n more questions all the reduced states are singlets or null states, then the n-state is called a **winning n-state**. If no such strategy exists, the state is called a **losing n-state**.

A state $(y_0, y_1, ..., y_e)$ is called a **substate** of $(x_0, x_1, ..., x_e)$ if $y_i \le x_i$ for each i. Clearly, any substate of a winning n-state is also a winning n-state.

When a state $(x_0, x_1, ..., x_e)$ is composed entirely of even integers, it is sensible to ask the question $\left[\frac{x_0}{2}, \frac{x_1}{2}, ..., \frac{x_e}{2} \right]$ to get two equal reduced states. We call this **bisecting**. More generally, for any state $(x_0, x_1, ..., x_e)$, we define **bisecting** to mean asking the question

$$\left[\left\lceil \frac{x_0}{2} \right\rceil, \left\lceil \frac{x_1}{2} \right\rceil, ..., \left\lceil \frac{x_e}{2} \right\rceil \right].$$

Lemma 2.1 (cf Hill and Karim (1992), Lemma 2.3). Bisecting the state $(2^m, 0, 0, ..., 0)$ m times gives the state $(1, m, \binom{m}{2}, ..., \binom{m}{e})$.

Example 2.2 We will prove the result of Pelc (1987) that 25 questions suffice to find one of 2^{20} objects in the 1-lie game.

By Lemma 2.1, bisecting $(2^{20}, 0)$ twenty times gives the state $(1,20)$. The following questioning strategy shows that $(1,20)$ is a winning 5-state. The obvious reductions of the winning 4-state $(0,16)$ and of the state $(0,4)$ have been omitted (these are states for a 0-lie game for which a simple bisecting strategy may be used).

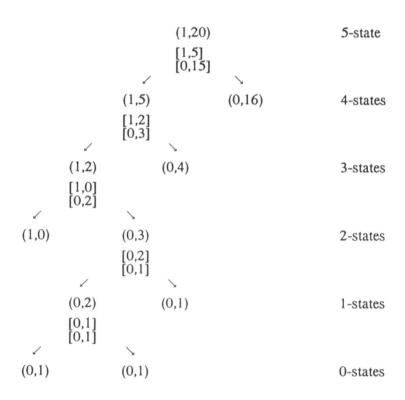

TWO BOUNDS

Berlekamp (1968) proved two important bounds: the **volume bound** (which has been rediscovered by several subsequent authors) and the **translation bound** (which has not).

The **volume** of an n-state $x = (x_0, x_1, ..., x_e)$ is defined by

$$V_n(x) = \sum_{i=0}^{e} x_i \sum_{j=0}^{e-i} \binom{n}{j} .$$

Intuitively, the volume of an n-state can be interpreted as the number of possibilities for the Responder if the Questioner still has n queries left; for each of the x_i objects in X_i (the set of possible targets with i lies already told), there are $\binom{n}{j}$ possibilities of lying exactly j times during the remaining series of n questions, and j can take any of the values from 0 to e - i.

It is an easy exercise (see Theorem 3.1 of Berlekamp (1968)) to show that conservation of volume holds for any question, i.e. if x is any n-state and if y and z are the (n-1)-states which result from asking any given question, then

$$V_n(x) = V_{n-1}(y) + V_{n-1}(z) .$$

Since the volume of a singlet 0-state is 1, it follows by conservation of volume and induction that:

Theorem 2.3 (The **volume bound**; Berlekamp (1968))
If x is a winning n-state, then $V_n(x) \le 2^n$.

Corollary 2.4 In particular, if the initial state (M,0,0,..., 0) is a winning n-state, then

$$M \sum_{j=0}^{e} \binom{n}{j} \le 2^n . \tag{2.1}$$

i.e. $f(M,e) \ge$ least value of n which satisfies (2.1).

Remarks

(1) The volume bound for the initial state is just the same as the Hamming, or sphere-packing, bound for one-way codes.

(2) By Corollary 2.4, $f(2^{20},1) \ge 25$. Hence Example 2.2 shows that $f(2^{20},1) = 25$.

The proof of the following bound is also basically by induction on n. However, it is less straightforward than the proof of the volume bound, and we refer to the proof of Theorem 3.4 in Berlekamp (1968) for the details.

Theorem 2.5 (The **translation bound**; Berlekamp (1968)).

Suppose $(x_0, x_1, ..., x_e)$ is a winning n-state with $\sum_{i=0}^{e} x_i \ge 3$. Then $(x_0, x_1, ..., x_{e-1})$ is a winning (n-3)-state.

Corollary 2.6 If $M \ge 3$, then $f(M,e) \ge f(M,e-1) + 3$. This says that for fixed $M(\ge 3)$, for each extra lie that may be told the minimum number of questions needed must go up by at least three.

INFINITE SEQUENCES OF WINNING STATES

The final ingredients needed for our solution of Ulam's problem are three tables of winning states constructed by Berlekamp (1968). The states meet the volume bound with equality. We give here, in Figure 2, only the most important of the three tables.

Let a_{ij} denote the i,j[th] entry of Figure 2, in which rows and columns are numbered 1, 2, 3, ... from the top and from the left, respectively. Let A_{im} denote the state $(a_{i1}, a_{i2}, ..., a_{im})$. An integer n beneath an entry a_{im} indicates that A_{im} is a winning n-state. So, for example, (1,4,14,58,246) is a winning 12-state.

The table is constructed as follows. The first two columns are postulated as initial conditions, the 1's and 0's continuing indefinitely. The remainder of the table is derived recursively by the following rules (applicable only when $j \geq 3$).

For $i \geq 3$, $a_{ij} = a_{i-1,j-1} + a_{i-2,j-1}$.

For $i = 2$, $a_{2j} = a_{3j} + a_{1,j-1}$.

For $i = 1$, $a_{1j} = a_{2j} + a_{3j}$.

From these Fibonacci-like relations, it can be shown that, for all $j \geq i$,

$$a_{ij} = 2 \left[\frac{1 + \sqrt{5}}{2} \right]^{3j-i-2} + 2 \left[\frac{1 - \sqrt{5}}{2} \right]^{3j-i-2}.$$

It follows also that:

Theorem 2.7

(i)   For $i = 1, 2$, the state A_{im} may be reduced to two identical copies of $A_{i+1,m}$. For $i \geq 3$, the state A_{im} may be reduced to the states $A_{i+1,m}$ and $A_{i-2,m-1}$.

(ii)   For $i \leq 2m$, A_{im} is a winning (3m-i)-state which meets the volume bound with equality.

(iii)   For fixed i, the state A_{im} becomes self-defining in the sense that each coefficient a_{ij} (for $j \geq i + 2$) is given by the recurrence relation

$a_{ij} = 4a_{i,j-1} + a_{i,j-2}$.

(iv)   for all $j \geq i$ (with the exceptions of a_{11}, a_{12} and a_{22}) a_{ij} is the nearest

Figure 2. An infinite sequence of winning states.

4	8	36	152	644	2728	11556	48952
2	*5*	*8*	*11*	*14*	*17*	*20*	*23*
2	6	22	94	398	1686	7142	30254
1	*4*	*7*	*10*	*13*	*16*	*19*	*22*
1	4	14	58	246	1042	4414	18698
	3	*6*	*9*	*12*	*15*	*18*	*21*
1	1	10	36	152	644	2728	11556
	2	*5*	*8*	*11*	*14*	*17*	*20*
1	0	5	24	94	398	1686	7142
		4	*7*	*10*	*13*	*16*	*19*
1	0	1	15	60	246	1042	4414
		3	*6*	*9*	*12*	*15*	*18*
1	0	0	6	39	154	644	2728
			5	*8*	*11*	*14*	*17*
1	0	0	1	21	99	400	1686
			4	*7*	*10*	*13*	*16*
1	0	0	0	7	60	253	1044
				6	*9*	*12*	*15*
1	0	0	0	1	28	159	653
				5	*8*	*11*	*14*
1	0	0	0	0	8	88	412
					7	*10*	*13*
1	0	0	0	0	1	36	247
					6	*9*	*12*
1	0	0	0	0	0	9	124
						8	*11*
1	0	0	0	0	0	1	45
						7	*10*
1	0	0	0	0	0	0	10
							9
1	0	0	0	0	0	0	1
							8

R. Hill

integer to

$$2\left(\frac{\sqrt{5}-1}{2}\right)^{i+2}\left(2+\sqrt{5}\right)^{j}.$$

(v)   Hence, for $j \geq i$, successive terms a_{ij} in the state A_{im} grow exponentially at the rate of $2 + \sqrt{5}$.

Remarks

(1)   The results in Theorem 2.7 are either proved in Berlekamp (1968) or are simple consequences of preceding parts.

(2)   The complete partitioning strategy from any state in Figure 2 is given by successive applications of Theorem 2.7(i). If one inserts arrows from each n-state in the diagram to the two reduced (n-1)-states (for each entry in the top two rows, insert a double arrow to the entry below; for each entry in the third and subsequent rows, insert one arrow to the entry below and the other arrow to the entry which is two places above and one place to the left), then one gets a flow-chart through which one can trace the complete state-reduction strategy from any given starting point.

For example, the state reduction from the 12-state (1,4,14,58,246) would begin as follows:

	(1,4,14,58,246)		12-state
⟋		⟍	
(1,1,10,36,152)		(4,8,36,152)	11-states
↓	⟍	↓↓	
(1,0,5,24,94)		(2,6,22,94)	10-states

3. THE SOLUTION OF ULAM'S PROBLEM FOR M = 2^{20} AND M = 10^6

The values of $f(2^{20},e)$ and $f(10^6,e)$ were given in Figure 1. In this section we shall summarise how these results were derived. The full details will be given in Hill, Karim and Berlekamp (to appear).

Lemma 3.1 $f(2^{20},e) \leq 3e + 26$, for all e.

Proof By Lemma 2.1, bisecting the initial state $(2^{20},0,0,...,0)$ twenty times gives the state $\left(1,20,190,1140,4845,...,\binom{20}{e}\right)$ whose i^{th} entry is the binomial coefficient $\binom{20}{i}$. Bisecting this a further six times gives the state

$$x = (1,1,5,41,233,1028,3597,10278,24411,48821,...)$$

(or substates of this).

By comparison with the state

$$y = (1,4,14,58,246,1042,4414,18698,...)$$

of Figure 2, we see that x is a winning 3e-state. (Note that, by Theorem 2.7(v), the entries in this last state are growing at the rate of approximately $2 + \sqrt{5}$. On the other hand, the entries of x are closely approximated by $\binom{26}{i}/2^6$ and so are growing at a slower rate than those of y from the sixth entry onwards; indeed the entries of x start to decrease from the thirteenth term onwards.)

So, in the e-lie game, $(2^{20},0,0,...,0)$ is a winning (3e+26)-state.

□

Theorem 3.2 $f(2^{20},e) = 3e + 26$ for e $\geq$ 8.

Proof By the volume bound (Corollary 2.4), $f(2^{20},8) \geq 50$. Hence, by the translation bound (Theorem 2.5), $f(2^{20},e) \geq 3e + 26$ for all e $\geq$ 8. This, combined with Lemma 3.1, gives the desired result. □

We have shown that for $e \geq 8$, the translation bound is the appropriate lower bound on $f(2^{20},e)$, and that this bound can always be achieved. For $e \leq 8$, it turns out that the volume bound is the appropriate lower bound and that this can always be achieved.

Theorem 3.3 (Hill, Karim and Berlekamp (to appear))

For $e \leq 7$, the values of $f(2^{20})$ are as shown in Figure 1.

Outline Proof   All of the proposed values are lower bounds by the volume bound (Corollary 2.4). So in each case it is sufficient to find an appropriate questioning strategy which achieves the bound.

For $e = 4$, after bisecting the initial state 20 times we reach the state $\underline{x} = (1,20,190,1140,4845)$. The desired result now follows immediately by comparing $\underline{x}$ with the winning 17-state $\underline{y} = (2,20,220,2511,28796)$ constructed in Figure 9 of Berlekamp (1968). Comparing successive truncations of $\underline{x}$ and $\underline{y}$ now gives the result for $e = 3, 2$ and 1.

For $e = 5$, the result follows by comparison of $(1,20,190,1140,4845,15504)$ with the winning 20-state $(6,22,196,1156,10954,43318)$ constructed in Figure 11 of Berlekamp (1968).

This leaves cases $e = 6$ and $e = 7$, which proved somewhat difficult. The problem is that $\left(1,20,...,\binom{20}{7}\right)$ is a 26-state which only just survives the volume bound. We eventually found a suitable questioning strategy, but we spare the reader the details here.

It seems perhaps paradoxical that for several months we knew how many questions were needed when a billion lies could be told (by Theorem 3.2, the answer is three billion and twenty-six, irrespective of whether we are talking about British or American billions!), but we could not say what the correct answer was for just six lies.

Let us now consider Ulam's problem for $M = 10^6$. Since $10^6 < 2^{20}$, we clearly have $f(10^6,e) \leq f(2^{20},e)$ for all e. Now all the lower bounds on $f(2^{20},e)$ which we found via the volume and translation bounds are easily shown also to be lower bounds on $f(10^6,e)$ with the one exception of $e = 4$, where the volume bound gives $f(2^{20},4) \geq 37$ but $f(10^6,4) \geq 36$. Again, a concerted attack was required to show that this last bound could be attained, and again we leave the details for elsewhere.

$\square$

4. ULAM'S PROBLEM FOR OTHER VALUES OF THE PARAMETERS

In this section we survey the known results concerning the function $f(M,e)$ for general M and e. First we consider the case where e is fixed and M varies, then the case where M is fixed but e varies, and finally we mention some asymptotic results.

4.1 A FIXED NUMBER e OF LIES

When e is fixed, it turns out that for sufficiently large M one can always achieve, or get very close to, the volume bound. To describe the known results, a useful concept is that of "survival", introduced by Spencer (1992). Let us first consider how an initial n-state (M,0,...,0) which satisfies the volume bound can fail to be a winning state. The problem is that if the volume bound is tight, then it will be necessary to split the volume very evenly at each question. Early in the questioning strategy, it may be hard to do this. For example, (3,0) is a 4-state which satisfies the volume bound (since $V_4((3,0)) = 15 \leq 2^4$) but is nevertheless a losing 4-state. The state cannot survive the first question without one of the reduced states then failing the volume bound.

Let us say that an n-state **survives** t questions if there exists a questioning strategy such that all the reduced (n-t)-states after t questions satisfy the volume bound.

We have just shown that the 4-state (3,0) cannot survive one question. On the other hand, suppose an n-state x with volume V is **balanced**, by which is meant that there exists a question such that the reduced (n-1)-states have volumes $\left\lfloor \dfrac{V}{2} \right\rfloor$ and $\left\lceil \dfrac{V}{2} \right\rceil$. If x satisfies the

volume bound (i.e. $V \leq 2^n$), then clearly the reduced states do so also
$\left(\text{since } \left\lceil \frac{V}{2} \right\rceil \leq 2^{n-1}\right).$

Thus, any balanced state can survive at least one question.

Now if an initial state is such that it can survive the early questions, then things become easier for the Questioner. The volume of a current state will become more concentrated to the right than to the left of the state and this makes an even splitting of the state easier to achieve. For example, any state $x = (x_0, x_1, ..., x_e)$ in which x_e is sufficiently large compared with the other x_i's will be balanced (because for a given question $[u_0, u_1, ..., u_e]$, if we change u_e by 1, then the volume of one reduced state goes up by 1 while the volume of the other goes down by 1).

Thus the first e questions are where the Questioner is most vulnerable. Once e questions have been asked, there will be a non-zero x_e term to help in splitting the subsequent volumes evenly.

Two further definitions, commonly used in several of the references, will be useful.

The **character** of a state x is defined as
$$ch(x) = \min \{n : V_n(x) \leq 2^n\} .$$
A state is called **nice** if it is a winning n-state for $n = ch(x)$.

Thus the initial state $(M,0,...,0)$ is nice if and only if $f(M,e) = ch((M,0,...,0))$. We saw above that the state $(3,0)$ is not nice. Example 2.2 shows that the state $(2^{20},0)$ is nice.

We are now ready to review the known results when e is fixed.

(1)   e = 1

As we saw in Section 1, the value of both $f(10^6,1)$ and $f(2^{20},1)$ was shown to be 25 by Pelc (1987). In fact, Pelc's paper solves Ulam's problem in the 1-lie game for all values of M, as follows.

Theorem 4.1 (Pelc (1987))

(a)   If M is even, then $f(M,1) = \min \{n : M(n+1) \leq 2^n\}$

(b)   If M is odd, then $f(M,1) = \min \{n : M(n+1) + n - 1 \leq 2^n\}$.

This is equivalent to saying that (a) if M is even, then the state $(M,0)$ is nice; (b) if M is odd; then the worst state, $\left(\frac{M+1}{2}, \frac{M-1}{2}\right)$ after the first question is nice.

In terms of survival, Pelc's result shows that for all M and n, the Questioner wins the (n,M,1) game if and only if he survives the first question.

(2) e = 2

For every case where M is a power of 2, Ulam's problem was solved for the 2-lie game by Czyzowicz, Mundici and Pelc (1989). They showed that $(2^m,0,0)$ is nice for all m except m = 2. In the exceptional case, f(4,2) = 8 = ch((4,0,0)) + 1.

Then Guzicki (1990) solved the 2-lie problem for all values of M. He showed that for M $\geq$ 90, the state (M,0,0) is a winning n-state if and only if a straightforward condition on M and n holds. The condition depends on the values of both M and n modulo 4, and we refer to Guzicki's paper for the details. The result essentially says that for M $\geq$ 90, the Questioner wins the (n,M,2) game if and only if he can survive the first two questions. In fact, inspection of the cases M < 90 shows that this last statement applies for all M $\geq$ 31. However, if M = 30 and n = ch((30,0,0)) = 11, the Questioner has no winning strategy even though he can survive the first two (and several more) questions.

For M $\leq$ 89, Guzicki showed, by individual case treatments, that except for the numbers 3, 4, 5, 6, 9, 10, 11, 17, 18, 29, 30, 51, 89, all other numbers are nice, i.e. the best strategy requires ch((M,0,0)) questions. The exceptional numbers require one more question.

A corollary of Guzicki's result is that in all cases, the questioner has a winning strategy in K or K + 1 questions where K is the character of (M,0,0).

For e = 3, Guzicki suggests that a complete analysis of the problem would be very much more difficult than for e = 2.

(3)   e = any fixed number

Ulam's searching game with a fixed number e of lies was considered for arbitrary e by Spencer (1992). He generalized the above results by showing that in the (n,M,e) game with fixed e, for sufficiently large M the Questioner has a winning strategy if and only if he can survive the first e questions.

4.2   A FIXED NUMBER M OF OBJECTS

The results of Section 4.1 imply that for fixed e, and sufficiently large M, the value of f(M,e) is essentially given by the smallest number of questions allowable by the volume bound, except that a very few additional questions may be needed in order to survive the first e questions.

However, if we consider a fixed number M of objects and let e get large, we find that the volume bound becomes no longer achievable. (It should be noted that M = 1 and M = 2 are trivial exceptional cases, where f(1,e) = 0 and f(2,e) = 2e + 1 for all e.)   As was shown by Berlekamp (1968), if M $\geq$ 3, the translation bound takes over from the volume bound as e gets large and it is found that f(M,e)/e $\to$ 3 as e $\to$ ∞. (If the volume bound could have been achieved for large e, we would have had f(M,e)/e $\to$ 2).

We have already seen this phenomenon illustrated for M = 2^{20} in Section 3. We saw that the volume bound can be achieved for e $\leq$ 8, but then the translation bound sets in with the value of f(2^{20},e) having to increase by 3 for each additional lie.   For e $\geq$ 8, we have f(2^{20},e)/e = (3e+26)/e $\to$ 3 as e $\to$ ∞ .

We have carried out a similar analysis to that in Section 3 for M = 2^h for all h $\leq$ 20, and have determined the values of f(2^h,e), for all e, in all cases except h = 14 and h = 19. The results are shown in Figure 3 and proofs will be in Karim (to appear). Many of the results for h $\leq$ 16 were also obtained independently by Lawler and Sarkissian (1993) by means of computer searches.

Figure 3   The values of $f(2^h, e)$ for $h \leq 20$

M	$f(M,e)$		M	$f(M,e)$	
2	2e + 1	for e ≥ 0	2^{11}	3e + 13	for e ≥ 2
2^2	3e + 2	for e ≥ 0	2^{12}	3e + 15	for e ≥ 4
2^3	3e + 3	for e ≥ 0	2^{13}	3e + 16	for e ≥ 3
2^4	3e + 4	for e ≥ 0	2^{14}	3e + (17or18)	for e ≥ 6
2^5	3e + 6	for e ≥ 1	2^{15}	3e + 19	for e ≥ 5
2^6	3e + 7	for e ≥ 1	2^{16}	3e + 20	for e ≥ 4
2^7	3e + 8	for e ≥ 1	2^{17}	3e + 22	for e ≥ 8
2^8	3e + 9	for e ≥ 1	2^{18}	3e + 23	for e ≥ 6
2^9	3e + 11	for e ≥ 2	2^{19}	3e + (24or25)	for e ≥ 8
2^{10}	3e + 12	for e ≥ 2	2^{20}	3e + 26	for e ≥ 8 .

For smaller values of e than those covered above, the volume bound is always attained.

For the exceptional cases, h = 14 or 19, at present we know the value of f(M,e) only to within 1 for large values of e.

For those values of e where the volume bound applies, finding a suitable questioning strategy is immediate in most cases (by a straightforward comparison with a state from Berlekamp's tables) but can be tricky when the volume bound is tight. We have already remarked on the difficulty of showing that $f(2^{20}, 7) = 46$. The problem is that the volume of the initial 46 - state $\underline{x}$ occupies over 96% of the "space" available; i.e. $V_{46}(\underline{x})/2^{46} \cong 0.9603$. An even tighter bound applies in showing that $f(2^{17}, 3) = 29$, because the initial 29-state occupies over 99.85% of the space available, which means that very even splittings are required right through the questioning strategy. However, the fact that e is as small as 3 makes a solution relatively easy to find and we pose this as an exercise for the reader.

4.3 ASYMPTOTIC RESULTS

For large values of M (the number of objects), n (the number of questions allowed) and e (the number of lies permitted), Berlekamp (1968) essentially solved the problem of when the Questioner has a winning strategy. Let R = (log₂M)/n be the **rate** and F = e/n be the **error fraction**. Then the Questioner has a winning strategy for those values of F lying below the curve shown in Figure 4.

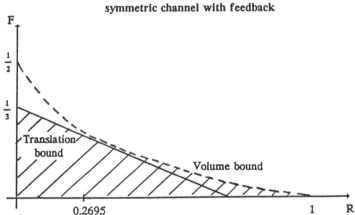

Figure 4. Asymptotic error-correction fraction of binary symmetric channel with feedback

For high rate codes, the volume bound (asymptotically this is the curve R = 1 - H₂(F), where H₂(F) = -F log₂ F - (1-F) log₂(1-F)) is the limiting bound on what can be achieved. But for low rate codes the translation bound gives the limiting bound. In particular, for "zero rate" codes, i.e. for fixed M as e → ∞, we find that f(M,e)/e → 3. So for fixed M (≥ 3) and large e we need to ask roughly 3 times as many questions as there are lies.

5.   ULAM'S PROBLEM WITHOUT FEEDBACK

Consider a non-interactive version of Ulam's game in which all the questions must be asked in advance, so that the Questioner cannot benefit from intermediate feedback. The Questioner collects all the answers and then tries to determine the unknown object.

Let $g(M,e)$ be the least number of questions in this non-interactive game which are sufficient to determine one of M objects if up to e of the answers may be lies. Clearly $g(M,e) \geq f(M,e)$.

Example 5.1   We shall show that $g(16,1) = 7$ with the aid of the following party trick. (This is a nice game to play with young children. You can convince them that you know exactly when they are lying!)

Ask someone (the Responder) to think of a number in the range 0 to 15. The Responder must give answers to the following seven questions, lying at most once.

Question 1.   Is your number one of 0, 1, 2, 3, 4, 5, 6, 7 ?
Question 2.   Is your number one of 0, 1, 2, 3, 8, 9, 10, 11, ?
Question 3.   Is your number one of 0, 1, 4, 5, 8, 9, 12, 13 ?
Question 4.   Is your number one of 0, 2, 4, 6, 8, 10, 12, 14 ?
Question 5.   Is your number one of 0, 3, 5, 6, 8, 11, 13, 14 ?
Question 6.   Is your number one of 0, 3, 4, 7, 9, 10, 13, 14 ?
Question 7.   Is your number one of 0, 2, 5, 7, 9, 11, 12, 14 ?

To illustrate how the Questioner determines the unknown number, suppose the sequence of answers is NNNYNNY (Y standing for yes and N for no). First note whether the number of N's in the last four positions is even or odd and mentally record a "0" or "1" accordingly. Now do the same thing for positions 2, 3, 6 and 7. Finally, do the same thing for the alternating positions 1, 3, 5 and 7. For the given example, the binary sequence 011 is thus formed. This is just the binary representation of the question number to which the Responder lied (000 would indicate no lies). The Questioner thus corrects the wrong answer and now finds the unknown number to be that whose binary representation is the first four (corrected) answers, with Y's identified with 0's and N's with 1's. So in this case the Questioner announces (almost instantly, with practice) that the third answer is a lie and the chosen number is 12.

How the trick works

Each number x ϵ {0,1,...,15} corresponds to a 7-bit codeword $\underline{x}$ as follows.

$$
\begin{array}{rcl}
0 & = & 0000000 \\
1 & = & 0001111 \\
2 & = & 0010110 \\
3 & = & 0011001 \\
4 & = & 0100101 \\
5 & = & 0101010 \\
6 & = & 0110011 \\
7 & = & 0111100 \\
8 & = & 1000011 \\
9 & = & 1001100 \\
10 & = & 1010101 \\
11 & = & 1011010 \\
12 & = & 1100110 \\
13 & = & 1101001 \\
14 & = & 1110000 \\
15 & = & 1111111 \, .
\end{array}
$$

This is just a Hamming code in which the first 4 bits of each codeword are the binary representation of the corresponding number. Note that asking Question i about the number x is the same as asking about the corresponding codeword $\underline{x}$: "Is the ith digit zero?" So the seven answers (with at most one lie), with yes written as 0 and no as 1, form the codeword $\underline{x}$ (with at most one error). The decoding algorithm is just the following standard syndrome decoding for a Hamming code. The codewords are precisely those vectors $\underline{x}$ such that $\underline{x} \, H^T = \underline{0}$, where

$$
H = \begin{bmatrix} 0 & 0 & 0 & 1 & 1 & 1 & 1 \\ 0 & 1 & 1 & 0 & 0 & 1 & 1 \\ 1 & 0 & 1 & 0 & 1 & 0 & 1 \end{bmatrix}
$$

is the check matrix of the code. Suppose the sequence of answers is $\underline{x} + \underline{e}$ where $\underline{e} = 0...010...0$ is the error vector with 1 in position j. Then the syndrome of the "answer vector" is $(x + \underline{e}) \, H^T = \underline{e} \, H^T =$ transpose of column j of H = binary representation of j.

This concludes our analysis of the party trick.

Note that $f(16,1) \geq 7$ by the volume bound and so we have $g(16,1)$ = $f(16,1) = 7$. This means that, in the case of 16 objects and one lie, we can do just as well without feedback as with!

More generally, we may note that any binary e-error correcting code of length n and size M may be used in the above way to give a strategy for determining one of M objects in n questions in the e-lie game. Conversely, it is easy to see that any winning strategy in the non-interactive e-lie game gives rise to a corresponding e-error correcting code. It thus follows, in general, that the value of $g(M,e)$ is simply the shortest length of a binary $(n,M,2e+1)$-code.

In the case of $M = 2^{20}$ and $e = 1$, we may use a $(25,2^{20},3)$ code, given by 6-times shortening a $(31,2^{26},3)$ Hamming code to show that $g(2^{20},1) = f(2^{20},1) = 25$. This observation was made by Niven (1988).

Finding $g(M,e)$ is a very difficult problem in general, bearing in mind that the code does not have to be linear. For example, the value of $g(2^{20},e)$ is unknown even for e as small as 2. Figure 5 gives the best known bounds on $g(2^{20},e)$ for e up to 10, and we see that the gap between the best known lower and upper bounds becomes wide fairly quickly as e increases. (For comparison, we list also the values of $f(2^{20},e)$ in Figure 5.)

Figure 5	Bounds on $g(2^{20},e)$	
e	$f(2^{20},e)$	$g(2^{20},e)$
0	20	20
1	25	25
2	29	29-30
3	33	33-35
4	37	37-40
5	40	40-43
6	43	43-52
7	46	46-54
8	50	50-61
9	53	53-65
10	56	56-70

The lower bounds on $g(2^{20},e)$ are just the values of $f(2^{20},e)$, while the upper bounds are given by the best known linear binary code as given by the tables of Brouwer and Verhoeff (1993).

On the basis of the examples discussed so far one might speculate on whether one can always do as well without feedback as with. For $M \leq 4$ one can indeed do so for all e. We leave it as a fairly easy exercise for the reader to show that

$$f(2,e) = g(2,e) = 2e + 1, \text{ for all e}$$
$$f(3,e) = g(3,e) = f(4,e) = g(4,e) = 3e + 2, \text{ for all e.}$$

But it is not always the case that $f(M,e) = g(M,e)$. For example $f(21,1) = 8$, while $g(21,1) = 9$. Also, we will next show (cf Berlekamp (1968)) that there is a clear distinction asymptotically between the two versions of the game, and that, for all $M > 4$, the values of $f(M,e)$ and $g(M,e)$ diverge as e increases. The Plotkin bound (cf page 315 of Berlekamp (1968A)) shows that any binary (n,M,d) code, i.e. a code of length n with M codewords and minimum Hamming distance d, satisfies

$$d \leq \frac{n}{2\left(1 - \frac{1}{M}\right)} .$$

Hence, putting $d = 2e + 1$, we have

$$g(M,e) \geq (4e+2)(1 - \frac{1}{M}) \qquad (5.1)$$

and so $$\lim_{e \to \infty} g(M,e)/e \geq 4(1 - \frac{1}{M}) . \qquad (5.2)$$

We next show that when M is a power of 2, the bound (5.1) can be attained, by a linear code, for infinitely many values of e.

Theorem 5.2 $g(2^h, t2^{h-2} - 1) = t2^h - t - 1$, for $t = 1, 2, ...$

Proof   Consider an $h \times (2^h - 1)$ matrix S(h) whose columns are the distinct non-zero binary vectors of length h. As is well known, S(h) is the generator matrix of a $(2^h - 1, 2^h, 2^{h-1})$ code called a **simplex code**.

Now consider a matrix obtained by placing t copies of S(h) side by side and then deleting any single column. This matrix clearly generates a $(t(2^h-1)-1, 2^h, t2^{h-1} - 1)$ code. From $d = t2^{h-1} - 1$, we have $e = (d-1)/2 = t2^{h-2} - 1$. Thus we have

$$g(2^h, t2^{h-2} - 1) \leq t2^h - t - 1.$$

On the other hand, by the Plotkin bound (5.1), we have

$$g(2^h, t2^{h-2} - 1) \geq \left\lceil (t2^h-2)(1-1/2^h) \right\rceil$$

$$= \left\lceil t2^h - t - 2 + 1/2^{h-1} \right\rceil$$

$$= t2^h - t - 1 \qquad \qquad \square$$

Let us return to the case of $M = 2^{20}$. Looking at Figure 5, one might expect that the gap between the best known lower and upper bounds would become hopelessly large as e increases. Theorem 5.2 shows that this is far from the case. The lower bounds in Figure 5 are all given by the volume bound, but as e gets large the dominant lower bound becomes the Plotkin bound which can be attained by the codes described in Theorem 5.2. So, for example, the continuation of Figure 5 would eventually reach entries such as

e	$f(2^{20},e)$	$g(2^{20},e)$
$2^{18} - 1$	3e + 26	4e + 2
$2^{19} - 1$	3e + 26	4e + 1
$2^{20} - 1$	3e + 26	4e - 1
$2^{21} - 1$	3e + 26	4e - 5 .

It seems remarkable that for 2^{20} objects, Ulam's problem without feedback is solved for the case of $2^{18} - 1$ lies but not for the case of 2 lies!

From the above observations, we see that for large M, as $e \to \infty$, the Questioner needs roughly 4 times as many questions as there are lies in the non-interactive game (as opposed to 3 times as many in the interactive game). In fact, the Questioner can sometimes do slightly better than this, asking roughly $4 - 1/M$ questions per lie. On the other hand, the Plotkin bound shows that for any fixed error fraction greater than $1/4$, for all sufficiently large values of M the Questioner has no winning strategy (as was observed by Spencer and Winkler (1992)).

Figure 6 summarizes the asymptotic situation for the non-interactive case. The Gilbert bound is a lower bound on the error fraction of the best coding strategy for given rate. Below this curve, the Questioner always has a winning strategy. The upper bound of McEliece et al is the best known asymptotic upper bound and the area between the two curves is a grey area where it is not known whether codes exist. Note that at "zero rate", for small M we can do better than appears to be indicated in Figure 6, since error fractions close to 1/4(1-1/M) can be achieved. This contrasts with the interactive case, in which it is known that, for any $M \geq 3$, no error fraction greater than $1/3$ can be achieved.

Figure 6.   Asymptotic error-correction fraction of binary symmetric channel without feedback.

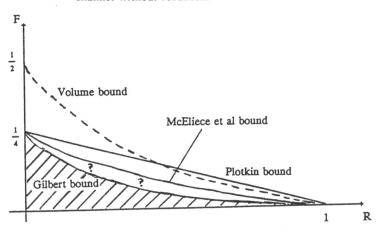

A comparison of Figures 4 and 6 shows that the best binary feedback coding strategies are asymptotically better than one-way binary block codes, at all rates R < 1 (the cases $M \leq 4$ are trivial exceptions).

6. OTHER VARIATIONS OF ULAM'S PROBLEM

6.1 A THIRD THRESHOLD

In Spencer and Winkler (1992), Ulam's problem is viewed from the following perspective. For a fixed error fraction $F = e/n$, the Questioner is said to **win** if for every M there exists a sufficiently large n (i.e. a sufficiently large rate) such that the Questioner has a winning strategy in the $(M,n,e = Fn)$ game. The authors show, as is suggested by earlier results in this article, that the threshold error fractions in Ulam's game with and without feedback are $1/3$ and $1/4$ respectively; more precisely, in the interactive game, the Questioner wins if $F < 1/3$ but loses if $F \geq 1/3$, while in the non-interactive game, the Questioner wins if $F \leq 1/4$ but loses if $F > 1/4$.

Spencer and Winkler (1992) consider also a third version of the game in which the questions are asked one at a time and the Responder is forbidden **at any point** to have lied to more than a fixed fraction F of them. So, for example, if $F < 1/3$, the first three answers must all be truthful. It turns out that the key threshold for this game is $1/2$; the Questioner wins if $F < 1/2$, but loses if $F \geq 1/2$. This result was obtained not only by Spencer and Winkler (1992) but also independently by Aslam and Dhagat (1991).

It is a most attractive feature of Ulam's problem that three very natural versions of the game result in the simple threshold fractions of $1/2$, $1/3$ and $1/4$.

6.2 ULAM'S GAME WITH COMPARISON QUESTIONS

Suppose that, in Ulam's game, only **comparison** questions may be asked about the unknown number x, i.e. questions of the form "Is $x \leq a$?" What is now the smallest number $f^c(M,e)$ of questions sufficient to determine a number in the set $\{1,2,..., M\}$ if up to e answers may be lies. This problem was considered in Rivest et al (1980), where some general bounds were obtained, in Spencer (1984), and in Aslam and Dhagat (1991).

Clearly $f^c(M,e) \geq f(M,e)$ and equality holds in the case $e = 0$. But for $e \geq 1$, it is not now as easy to split the volume of a current state evenly.

Example   We know from Example 5.1 that $f(16,1) = 7$. The 7-state $(16,0)$ meets the volume bound with equality and so exact halvings of the volume are required for every question in order to win in 7 questions. With comparison only questions, an exact halving is possible for the first question (Is $x \leq 8$?), giving the 6-state $(8,8)$, but the next comparison question cannot give an exact splitting and must now give a 5-state which violates the volume bound. Thus $f^c(16,1) \geq 8$. It is not hard to show that $f^c(16,1) = 8$.

Spencer (1984) showed that in the 1-lie comparison game, any state can be split sufficiently evenly that the Questioner can always win in at most $ch((M,0)) + 1$ questions. Thus $f^c(M,1) = ch((M,0))$ or $ch((M,0)) + 1$, for all M. With regard to Ulam's original problem, it remains an open problem (to the best of the author's knowledge) whether the value of $f^c(10^6,1)$ is 25 or 26.

6.3   DETECTING ERRORS IN SEARCHING GAMES

We have already observed that Ulam's problem is the interactive counterpart of finding the shortest length of a binary e-error **correcting** code of size M. Pelc (1989) considered a problem which is the interactive counterpart of finding a shortest e-error **detecting** code of size M. Ulam's game is modified by keeping the same rules except for a new definition of the Questioner's win; now he wins if either he finds the unknown x or if he can prove that the Responder lied at least once (possibly without being able to tell when). Pelc gives a simple proof, which we leave as an exercise for the reader, that the minimum number of questions sufficient to win the e-lie game with M objects is

$\lceil \log_2 M \rceil + e$.

In the same paper, Pelc considers a game with the same rules except that the Questioner now wins if either he finds the unknown x or he can tell **how many** times the Responder lied. (The non-interactive counterpart is to find a code of shortest length having minimum distance 2e).

Let $f^d(M,e)$ denote the minimum number of questions sufficient to win this game. Pelc solves the problem for e = 2, showing that

$$f^d(M,2) = \lceil \log_2 M \rceil + 3$$

and shows in general that

$$\lceil \log_2 M \rceil + e \leq f^d(M,e) \leq \lceil \log_2 M \rceil + 3e.$$

6.4   q-ARY SEARCHING WITH LIES

Malinowski (1994) considered the problem of determining the minimum number $f_q(M,e)$ of questions sufficient to find an unknown integer x between 1 and M in an e-lie game if the admissible form of question is "To which one of the disjoint sets A_1, A_2, ..., A_q does x belong?" Malinowski solves the problem for e = 1 for all M and q (the case e = 1 and q = 2 was that already solved by Pelc (1987)).

An **n-state** $\underline{x} = (x_0, x_1)$ in the q-ary 1-lie game is defined as for q = 2, and the **volume** $V_n(\underline{x})$ of $\underline{x}$ is defined to be $((q-1)n + 1)x_0 + x_1$. The **volume bound** is now $V_n(\underline{x}) \leq q^n$.

Malinowski shows that for q = 3, as for q = 2, the Questioner wins the (n,M,1) game if and only if he survives the first question (i.e. if and only if the worst state, after asking the most sensible first question, is an (n-1)-state satisfying the volume bound). For general q, the situation is a little more complicated; the Questioner always has a winning strategy subject to surviving the first question in the case where $M \geq q^{q-1}$, while a precise formula for $f_q(M,1)$ is given also for smaller values of M.

As for the binary case, we may also consider the problem of finding the smallest number $g_q(M,e)$ of questions sufficient to win the q-ary e-lie game without feedback. Again, $g_q(M,e)$ is just the smallest length of a q-ary (n,M,2e+1) error correcting code. So, for example, in the 1-lie game, q-ary Hamming codes show that $g_q(q^{n-r},1) = f_q(q^{n-r},1) = n,$ whenever n is an integer of the form $(q^r-1)/(q-1)$.

6.5   DETECTING A COUNTERFEIT COIN WITH UNRELIABLE WEIGHINGS

Consider the problem of determining the minimum number of weighings on a balance to find a counterfeit (heavier) coin x from among M coins $\{1,2,...,M\}$ if at most one weighing result may be erroneous.

Each question admits three possible answers: left pan goes down (which means x is among the coins on the left pan), right pan goes down (x is on the right pan), the pans are balanced (x is among the remaining coins). This is a 3-ary searching problem in the sense of section 6.4, except that questions are restricted to those in which two of the three sets we ask about must have equal sizes (the numbers of coins on left and right pan must be equal). This problem was solved by Pelc (1989A).

6.6 "MASTERMIND" WITH LIES

We leave a final challenge for the reader. in the standard game of Mastermind, the Responder chooses an ordered 4-tuple of colours from a 6-set. The Questioner tries to determine this by asking questions, each question taking the form of a guess at the unknown 4-tuple. The Responder's answer is to say how many of the entries in the guess are correct (i.e. the right colour in the right position) and also how many other colours are correct but in the wrong position. It was shown by Knuth (1976) that the Questioner can always win in 5 guesses. But now suppose the Responder is allowed one lie? How many questions are now needed to guarantee a win for the Questioner?

REFERENCES

Aslam, J A and Dhagat, A. (1991). Searching in the presence of linearly bounded errors. In Proceedings of the 23rd Annual ACM Symposium of Theory of Computing.

Berlekamp, E R (1964). Block coding with noiseless feedback, PhD thesis, Department of Electrical Engineering, Massachusetts Institute of Technology, Cambridge, Massachusetts.

Berlekamp, E R (1968). Block coding for the binary symmetric channel with noiseless, delayless feedback, in *Error-correcting Codes*, edited by H B Mann, Wiley, New York, pp 61-88.

Berlekamp, E R (1968A). *Algebraic Coding Theory*, McGraw-Hill, New York.

Brouwer, A E and Verhoeff, T (1993). An updated table of minimum distance bounds for binary linear codes, *IEEE Trans Inform Theory* IT **39**, 662-677.

Czyzowicz, J, Mundici, D and Pelc A. (1988). Solution of Ulam's problem on binary search with two lies. *J Combin Theory* Ser A **49**, 384-388.

Czyzowicz, J, Mundici, D and Pelc, A. (1989). Ulam's searching game with lies. *J Combin Theory* Ser A **52**, 62-76.

Guzicki, W. (1990). Ulam's searching game with two lies. *J Combin Theory* Ser A **54**, 1-19.

Hill, R and Karim, J P (1992). Searching with lies: the Ulam problem. *Discrete Math.* **106/107**, 272-283.

Hill, R, Karim, J P and Berlekamp, E R (to appear). The solution of Ulam's problem on searching with lies.

Karim, J P (to appear), PhD thesis, Department of Mathematics and Computer Science, University of Salford.

Knuth, D E (1976-77). The Computer as Mastermind. *J Recreational Math* **9**, 1-6.

Lawler, E L and Sarkissian, S. (1993). Adaptive error-correcting codes based on co-operative play of the game of "twenty questions with a liar", preprint.

Malinowski, A. (1994). K-ary searching with a lie. *Ars Combinatoria* **37**, 301-308.

Negro, A and Sereno, M. (1992). Solution of Ulam's problem on binary search with three lies. *J Combin Theory* Ser A **59**, 149-154.

Niven, I. (1988). Coding theory applied to a problem of Ulam. *Math Mag.* **61**, 275-281.

Pelc, A. (1987). Solution of Ulam's problem on searching with a lie. *J Combin Theory* Ser A **44**, 129-140.

Pelc, A. (1989). Detecting errors in searching games. *J Combin Theory* Ser A **51**, 43-54.

Pelc, A. (1989A). Detecting a counterfeit coin with unreliable weighings, *Ars Combinatoria* **27**, 181-192.

Rivest, R L, Meyer, A R, Kleitman, D L, Winklmann, K and Spencer, J. (1980). Coping with errors in binary search procedures. *J Comput System Sci* **20**, 396-404.

Spencer, J. (1984). Guess a number - with lying. *Math Mag* **57**, 105-108.

Spencer, J. (1992). Ulam's searching game with a fixed number of lies. *Theoretical Computer Science* **95**, 307-321.

Spencer, J and Winkler, P. (1992). Three thresholds for a liar. *Combinatorics, Probability and Computing* **1**, 81-93.

Ulam, S M. (1976). *Adventures of a Mathematician*, Scribner, New York, p 281.

SPIN MODELS FOR LINK INVARIANTS

François Jaeger
Laboratoire de Structures Discrètes et de Didactique,
IMAG, GRENOBLE

1. INTRODUCTION

The discovery by Vaughan Jones in 1984 of a new polynomial invariant of links was the starting point of spectacular advances in knot theory which suddenly brought together previously unrelated concepts from various branches of mathematics and physics. One particularly fruitful idea was to consider a link diagram as an abstraction of a physical system of elementary objects (molecules, atoms, particles...) interacting in a local fashion. These local interactions are described by a *statistical mechanical model*. In the context of physics, much of the relevant information is then given by the *partition function*. The basic facts from the point of view of knot theory are that one can define natural conditions on the parameters of the model which insure that the partition function is a link invariant, and that one can actually find models satisfying these conditions which yield non-trivial link invariants. In fact, the models which correspond to link invariants are closely related to the *exactly solvable* or *integrable* models which are of particular interest in physics.

We shall be mainly concerned here with the version of this approach to the construction of link invariants which is based on *spin models*. There the local interactions can be viewed as taking place between the vertices of a graph and along the edges of this graph. The simplest case is that of the *Potts models*, which give rise in the context of graph theory to the *Tutte polynomial*, and in the context of knot theory to the above mentioned *Jones polynomial*. Spin models for link invariants can be defined in terms of matrices satisfying certain equations. It turns out that these equations imply that each solution must be closely related to some *association scheme*. This leads to new relations between knot theory, graph theory and algebraic combinatorics.

71

Our purpose will be to survey this relatively recent topic, placing emphasis on algebraic and combinatorial aspects. In particular, we shall not examine spin models in the context of statistical mechanics; the interested reader can refer for instance to [Ba], [Bi2], [T]. Similarly, the description of the knot theory background will be reduced to a minimum. Also, we have tried to give a synthetic view which incorporates the most recent results, and in some cases this is at the expense of a chronological perspective. Finally, the present exposition significantly reflects personal interests and can claim neither objectivity nor completeness. Other surveys can be found in [B2], [H], [Ja2].

2. SPIN MODELS FOR GRAPHS AND FOR LINK DIAGRAMS

2.1. Spin models for graphs

All graphs will be finite, and loops and multiple edges will be allowed. Graphs will be directed, unless otherwise specified. For a given finite set X, n denotes the size of X, and M_X denotes the set of n×n complex matrices with rows and columns indexed by X.

Let G be a graph and w be a mapping from its edge-set to M_X whose values will be called *edge weights*. We shall call the pair (G,w) a *weighted graph*. Let us call *state of G* any mapping σ from the vertex-set of G to X. If the edge e has initial end v_1 and terminal end v_2, the *weight* w(e|σ) *of e with respect to the state σ* is the $(\sigma(v_1),\sigma(v_2))$-entry of the matrix w(e). The *weight* w(σ) *of the state σ* is then the product of the w(e|σ) over all edges e (this will be set to 1 if G has no edge). Finally, let Z(G,w) be the sum of weights of all states. Z(G,w) is the *partition function of the weighted graph (G,w)*.

Loosely speaking, a *spin model* is a specific way to assign edge-weights to graphs. This concept leads to interesting invariants of graphs. For instance if w(e) equals J - I for all edges e (where I is the identity matrix and J is the all-one matrix), Z(G,w) is clearly the number of proper vertex colorings of G with colors in X. A natural generalization is to replace J - I by an arbitrary linear combination of I and J. This gives (up to a normalization factor) Tutte's dichromatic polynomial (see [Tut]) and corresponds to the Potts model in statistical mechanics (see for instance [Ba], [Bi2]). Another natural generalization is to replace J - I by the adjacency matrix of some graph H, and then Z(G,w) counts the homomorphisms from G to H, considered as mappings between vertex sets (see [HJa]). Other examples appear in [HJo], which also gives a nice introduction to the application of statistical mechanical models to the construction of invariants of links, as initiated by V.F.R. Jones in [Jo3].

2.2. Spin models for link diagrams

A *link* consists of a finite collection of disjoint simple closed curves (the *components* of the link) smoothly embedded in 3-space. If each component has received an orientation, the link is said to be *oriented*. (Oriented) links can be represented by *(oriented) diagrams*. A diagram of a link is a generic plane projection (there is only a finite number of multiple points, each of which is a simple crossing), together with an indication at each crossing of which part of the link goes over the other. Such a diagram can be viewed as a 4-regular plane graph together with some additional binary information at each vertex (here we must allow *free loops*, i.e. edges without end-vertices represented by simple closed curves disjoint from the rest of the graph). Diagrams are considered up to plane deformations. For oriented links, the edges of the diagrams are oriented according to the orientations of the corresponding link components. The *Tait number* (or *writhe*) T(L) of an oriented diagram L is the sum of signs of its crossings, where the sign of a crossing is defined on Figure 1.

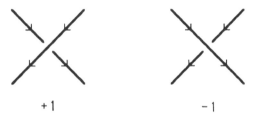

Figure 1

François Jaeger

Two links are *ambient isotopic* if there exists an isotopy of the ambient 3-space which carries one onto the other (for oriented links, this isotopy must preserve the orientations). This natural equivalence of links is described in terms of diagrams by Reidemeister's Theorem, which asserts that two diagrams represent ambient isotopic links if and only if one can be obtained from the other by a finite sequence of elementary local transformations, the *Reidemeister moves*. These moves belong to three basic types described for the unoriented case in Figure 2.

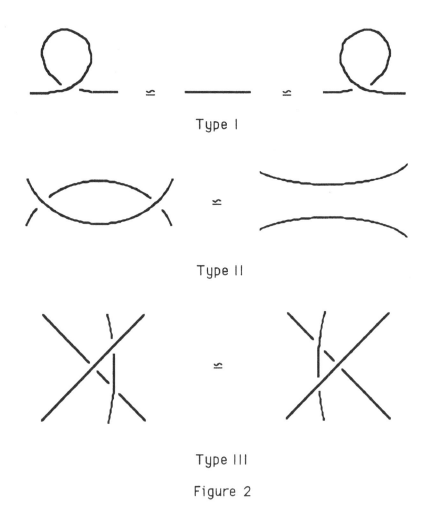

Type I

Type II

Type III

Figure 2

A move is performed by replacing a part of diagram which is one of the configurations of Figure 2 by an equivalent configuration without modifying the remaining part of the diagram. For the oriented case, all local orientations of these pairs of equivalent configurations must be considered. However some moves can be replaced by sequences of other moves and thus important simplifications are possible. More details can be found for instance in [BZ] or [K1].

Reidemeister's Theorem allows the combinatorial definition of a *link invariant* as an assignment of values to diagrams such that the value of any diagram is preserved by Reidemeister moves. As shown in [Jo3], one may use the partition functions of spin models to obtain such invariants. In [Jo3] the construction was restricted to spin models involving two symmetric edge weights, and the possibility of an extension to non-symmetric edge weights was suggested. This was carried out by K. Kawagoe, A. Munemasa and Y. Watatani [KMW]. An even more general construction involving four non-symmetric edge weights was finally introduced by Eiichi and Etsuko Bannai [BB4]. We now present these constructions, starting with the most general one.

With every connected diagram L we associate a plane graph G(L) as follows. The plane regions delimited by the diagram are colored with two colors, black and white, in such a way that adjacent regions receive different colors and the infinite region is colored white. Then G(L) has one vertex inside each black region (let us call it the capital of this region), and one edge through each crossing. Each edge is a simple curve which joins the capitals of the black regions incident to the corresponding crossing. This is done in such a way as to obtain a plane embedding of G(L) (this construction is classical in knot theory, see for instance [BZ]). Then spin models on a connected link diagram L will be defined as spin models on the associated graph G(L).

Assume now that L is oriented. Essentially four distinct situations can occur at a crossing. For each of these, we choose an orientation of the corresponding edge e of G(L) (according to the orientation of the upper part of the link), and an edge weight w(e) among four given matrices W_1, W_2, W_3, W_4 in M_X, where X is some set of size $n \geq 2$ (see Figure 3).

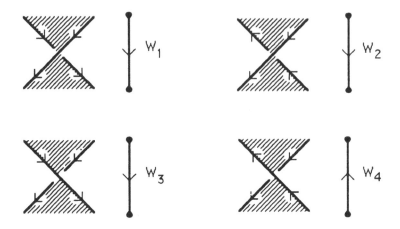

Figure 3

This is the definition of the edge weights for the *4-weight models* of [BB4]. If W_1, W_2 are equal to some matrix W^+ and W_3, W_4 are equal to some matrix W^-, we obtain the definition of [KMW]. For convenience we shall call these models *2-weight models* (although, as shown in [BB4], other interesting types of models involving only two edge weights exist). Finally, if the matrices W^+, W^- defining a 2-weight model are symmetric, we have the *symmetric 2-weight models* defined in [Jo3].

2.3. The invariance equations

We now look for natural conditions under which the partition function $Z(L) = Z(G(L),w)$, where w is defined in the previous section, will give an invariant of links. We shall assume that all diagrams are connected. This is not a significant restriction since every link can be represented by a connected diagram, and in Reidemeister's Theorem we may assume that each move preserves connectedness.

To illustrate the basic idea, let us study an example of Reidemeister move of type II for 2-weight models. There are two cases of local black and white coloring of the regions (see Figure 4). We shall consider a restricted set of "specified" states which have prescribed values on (the capital of) every black region, except the digon region destroyed by the move in the second case.

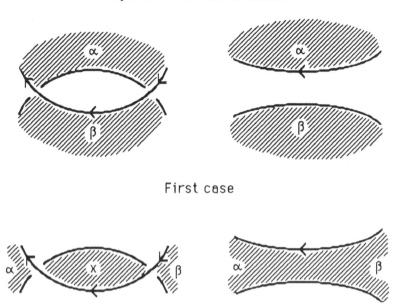

First case

Second case

Figure 4

We would like that for each such prescription, the sum of weights of specified states be the same for the two diagrams. Note that the weights of the edges corresponding to crossings which do not appear in the move are the same with respect to all specified states, and are also the same for the two diagrams. Factoring out these weights leads to the conditions:

(IIa) for every α, β in X, $W^+[\alpha,\beta]\, W^-[\beta,\alpha] = 1$.

(IIb) for every α, β in X, $\sum_{x \in X} W^+[x,\alpha]\, W^-[\beta,x] = \delta(\alpha,\beta)$, where δ is the Kronecker symbol.

These conditions are clearly incompatible (take $\alpha = \beta$ in (IIb)). This forces us to introduce some *normalization factor* in the partition function to compensate the difference between the situations (IIa) and (IIb). This normalization factor will be $D^{-|V(G(L))|}$, where D is a square root of $|X| = n$ (see [Jo3]). We shall replace the condition (IIb) by the condition

(IIc) for every α, β in X, $\sum_{x \in X} W^+[x,\alpha]\, W^-[\beta,x] = n\, \delta(\alpha,\beta)$,

which is now compatible with (IIa).

Then (IIa) and (IIc) will together imply that the quantity $D^{-|V(G(L))|} Z(L)$ is invariant under Reidemeister moves of type II. Note that for condition (IIc) we use our connectivity assumption, which implies that the simplest diagram has exactly two black regions less than the other.

The study of the Reidemeister move of type III where all arrows are directed downwards then leads similarly to the following *star - triangle equations* :

(IIIa) for every α, β, γ in X,

$$\sum_{x \in X} W^+[\alpha,x]\, W^+[x,\beta]\, W^-[x,\gamma] = D\, W^+[\alpha,\beta]\, W^-[\alpha,\gamma]\, W^-[\beta,\gamma] ;$$

(IIIb) for every α, β, γ in X,

$$\sum_{x \in X} W^+[\alpha,x]\, W^+[x,\beta]\, W^-[\gamma,x] = D\, W^+[\alpha,\beta]\, W^-[\gamma,\alpha]\, W^-[\gamma,\beta].$$

However the second equation (IIIb) can be derived from the first one (IIIa) together with (IIa), (IIc) (see [KMW]). Other oriented versions of the Reidemeister move of type III need not be considered since they can be replaced by sequences of moves of type II together with the already chosen move of type III (see for instance [Tur]). One can also derive algebraically from (IIa), (IIc), (IIIa) all other equations arising from Reidemeister moves of type III (see [KMW]).

Taking $\gamma = \alpha$ in (IIIa) and using (IIa), we obtain that for every α, β in X,

$$\sum_{x \in X} W^+[x,\beta]\ = D\, W^-[\alpha,\alpha] .$$

The same method applied to (IIIb) yields that for every α, β in X,

$$\sum_{x \in X} W^+[\alpha,x]\ = D\ W^-[\beta,\beta].$$

Thus there is a non-zero complex number a such that
(Ia) for every α in X,

$$\sum_{x \in X} W^+[x,\alpha]\ = \sum_{x \in X} W^+[\alpha,x] = D\, a^{-1} ;$$

(Ib) for every α in X, $W^-[\alpha,\alpha] = a^{-1}$.

We may rewrite (IIc) as the matrix equality $W^- W^+ = n\, I$, and similarly (Ia) as the equalities $J\, W^+ = W^+ J = D\, a^{-1} J$. Combining these equalities we obtain that $J\, W^- = W^- J = D\, a\, J$ or, equivalently,

(Ic) for every α in X,

$$\sum_{x \in X} W^-[x,\alpha]\ = \sum_{x \in X} W^-[\alpha,x] = D\, a.$$

Also it is clear from (IIa) and (Ib) that
(Id) for every α in X, $W^+[\alpha,\alpha]\ =\ a$.

For reasons which will become clear later, it is convenient to reformulate also equations (IIa), (Ib) and (Id) in matrix terms. Recall that the Hadamard product of two matrices A, B in M_X is denoted by $A \circ B$ and given by $(A \circ B)[x,y] = A[x,y] \, B[x,y]$ for all x, y in X. Then (IIa), (Ib) and (Id) are equivalent to the equalities $W^+ \circ W^{-T} = J$, $I \circ W^- = a^{-1}I$, and $I \circ W^+ = a \, I$.

Now a study of Reidemeister moves of type I easily shows that properties (Ia), (Ib), (Ic), (Id) imply the invariance of the quantity $Z'(L) = a^{-T(L)} D^{-IV(G(L))I} Z(L)$ under these moves. Since the Tait number $T(L)$ is invariant under Reidemeister moves of type II and III, $Z'(L)$ is invariant under all types of Reidemeister moves.

We may sum up the above discussion as follows.

We shall call a *2-weight spin model* a triple (X, W^+, W^-), where W^+, W^- are matrices in M_X which satisfy the following properties (1) to (5) for some numbers $a \neq 0$ and D with $D^2 = n = |X|$:

(1) $I \circ W^+ = a \, I,\ I \circ W^- = a^{-1} \, I$;

(2) $J \, W^+ = W^+ J = D \, a^{-1} J,\ J \, W^- = W^- J = D \, a \, J$;

(3) $W^+ W^- = n \, I$;

(4) $W^+ \circ W^{-T} = J$;

(5) (Star - triangle equation) for every α, β, γ in X,
$$\sum_{x \in X} W^+[\alpha,x] \, W^+[x,\beta] \, W^-[x,\gamma] = D \, W^+[\alpha,\beta] \, W^-[\alpha,\gamma] \, W^-[\beta,\gamma] \, .$$

We have sketched a proof of the result of [KMW] that to every 2-weight spin model corresponds a link invariant given for any connected link diagram L by the normalized partition function $Z'(L)$.

When the matrices W^+, W^- are symmetric, the spin model is said to be *symmetric* and the definitions and results of [Jo3] are recovered. An important feature of these symmetric models is of course that we do not need the orientations of L and G(L) in the definition of the partition function.

The situation for 4-weight spin models is more complicated but the basic ideas are similar. The definition of [BB4] is essentially the following. A *4-weight spin model* is a 5-tuple (X, W_1, W_2, W_3, W_4), where W_1, W_2, W_3, W_4 are matrices in M_X which satisfy the following properties (6) to (10) for some numbers $a \neq 0$ and D with $D^2 = n = |X|$:

(6) $IoW_1 = a\,I$, $IoW_3 = a^{-1}I$;

(7) $J\,W_2 = W_2\,J = Da^{-1}J$, $J\,W_4 = W_4\,J = Da\,J$;

(8) $W_1\,W_3 = n\,I = W_2\,W_4$;

(9) $W_1 o\,W_3{}^T = J = W_2 o\,W_4{}^T$;

(10) (Star - triangle equations) for every α, β, γ in X,

$$\sum_{x \in X} W_1[\alpha,x]\,W_1[x,\beta]\,W_4[\gamma,x] = D\,W_1[\alpha,\beta]\,W_4[\gamma,\alpha]\,W_4[\gamma,\beta],$$

$$\sum_{x \in X} W_1[x,\alpha]\,W_1[\beta,x]\,W_4[x,\gamma] = D\,W_1[\beta,\alpha]\,W_4[\alpha,\gamma]\,W_4[\beta,\gamma] .$$

Then again the normalized partition function $Z'(L)$ defines a link invariant [BB4]. It is also shown in [BB4] that, if $W_1 = W_2 = W^+$ and $W_3 = W_4 = W^-$, conditions (6)-(10) are equivalent to conditions (1)-(5).

We can now formulate two basic problems: classify 2-weight and 4-weight spin models, and describe as accurately as possible the corresponding invariants of links. The second problem asks in particular for a better understanding of link invariants coming from spin models in the context of knot theory, but also contains for instance the question of how to compute these invariants with reasonable efficiency. Here we shall mainly address the first problem and we shall see that the theory of association schemes plays a crucial role in its study.

3. ASSOCIATION SCHEMES : AN INTRODUCTION

We shall need the following basic facts concerning association schemes (see [BI], [D], [BCN] for more details).

A *d-class association scheme* on the finite non-empty set X is a partition of X x X into d+1 non-empty relations R_i , i =0,...d, where $R_0 = \{(x,x) / x \in X\}$, which satisfies the following properties:

(i) For every scheme relation R_i, $\{(y,x) / (x,y) \in R_i\}$ is also a scheme relation; that is, there exists j in $\{0,...d\}$ such that $\{(y,x) / (x,y) \in R_i\} = R_j$.

(ii) For x, y in X, the number of elements z which satisfy given scheme relations with x and y only depends on which scheme relation is satisfied by the pair (x,y). That is, for every i, j, k in $\{0,...d\}$ there exists an integer $p_{ij}{}^k$ (called an *intersection number*) such that $|\{z \in X / (x,z) \in R_i, (z,y) \in R_j\}| = p_{ij}{}^k$ for every x, y in X with (x,y) in R_k.

All association schemes considered here are *commutative*, which means that

(iii) $p_{ji}{}^k = p_{ij}{}^k$ for all i, j, k in {0,...d}.

Define matrices A_i , i=0,...d, in M_X by

(iv) $A_i[x,y]$ equals 1 if $(x,y) \in R_i$, and equals 0 otherwise.

The above properties can then be reformulated as follows:

(11) $A_i \neq 0$, $A_i \circ A_j = \delta(i,j) A_i$;

(12) $A_0 = I$;

(13) $\sum_{i=0,...,d} A_i = J$;

(14) $A_i{}^T = A_j$ for some j in {0,...d} ;

(15) $A_i A_j = A_j A_i = \sum_{k=0,...,d} p_{ij}{}^k A_k$.

Let **A** be the subspace of M_X spanned by the matrices A_i , i=0,...d. By (11) these matrices are linearly independent and hence form a basis of **A**. Then (11) and (13) imply that, under Hadamard product, **A** is an associative commutative algebra with unit J, and {A_i , i∈{0,...d}} is a basis of orthogonal idempotents of this algebra. Moreover by (14) **A** is closed under transposition. Finally it follows from (12) and (15) that under ordinary matrix product **A** is also an associative commutative algebra with unit I. The subspace **A** of M_X endowed with these two algebra structures is called the *Bose-Mesner algebra* [BM] of the association scheme. Conversely, a (d+1)- dimensional vector subspace of M_X which contains I, J, is closed under transposition, Hadamard product and ordinary matrix product, and for which the ordinary matrix product is commutative, is the Bose-Mesner algebra of some commutative d-class association scheme (and will be called here a *Bose-Mesner algebra on* X). The main reason is that a space of matrices closed under Hadamard product has a basis of orthogonal idempotents for this product, i.e. a basis of disjoint (0,1) matrices (see [BCN], Th. 2.6.1, whose proof is easily extended to the non-symmetric case). In general we shall work here with Bose-Mesner algebras rather than with the equivalent combinatorial concept of association scheme.

A Bose-Mesner algebra is *symmetric* if it consists only of symmetric matrices. A *duality* of a Bose-Mesner algebra **A** on a set X of size n is a linear map Ψ from **A** to itself which satisfies the following properties :

(16) For every matrix M in **A**, $\Psi(\Psi(M)) = n\, M^T$;

(17) For any two matrices M, N in **A**, $\Psi(MN) = \Psi(M) \circ \Psi(N)$.

It easily follows that

(18) For any two matrices M, N in **A**, $\Psi(M \circ N) = n^{-1}\, \Psi(M)\, \Psi(N)$;

(19) $\Psi(I) = J$;

(20) $\Psi(J) = n\, I$.

A Bose-Mesner algebra will be said to be *self-dual* if it admits a duality.

Classical results in linear algebra show that a Bose-Mesner algebra **A** on a set X of size n has a basis $\{E_i,\ i = 0,...d\}$ of orthogonal idempotents for the ordinary matrix product, where $E_0 = n^{-1}\, J$ (see [BI] Section II.3).

The *eigenmatrix* P relates the two bases of idempotents as follows:

(21) $A_j = \Sigma_{i=0,...,d}\, P[i,j]\, E_i$.

If Ψ is a duality of **A**, the matrices $\Psi(E_i)$, i = 0,...d, are the Hadamard idempotents A_i in some order. We may choose the indexes in such a way that $\Psi(E_i) = A_i$ for i = 0,...d. Then by (21) P is the matrix of Ψ with respect to the basis $\{E_i,\ i=0,...d\}$, and, by (16), $n^{-1}P^2$ is the matrix of the transposition map with respect to the same basis (which is the identity if **A** is symmetric).

4. EXAMPLES OF SPIN MODELS

4.1. Spin models for the Jones polynomial

The simplest example of a spin model which yields a link invariant appears in [Jo3] and is obtained by considering 2-weight models (X,W^+,W^-) such that W^+, W^- are linear combinations of I and J (in the terminology of statistical mechanics this is a *Potts model*). The following exposition is similar to section 4 in [H]. See also [HJo], [Ja1].

Write as before n = |X|. Equations (1), (4) show that $W^+ = aI + b\, (J - I)$ and $W^- = a^{-1}I + b^{-1}\, (J - I)$ for some non-zero complex number b. Equations (2) then give $a + (n-1)\, b = D\, a^{-1}$, $a^{-1} + (n-1)\, b^{-1} = D\, a$. It easily follows (since n > 1) that $a = b + D\, b^{-1}$, $a^{-1} = b^{-1} + D\, b$. This yields $D = -b^2 - b^{-2}$ and $a = -b^{-3}$.

It is now immediate to check algebraically that equation (3) holds.

Finally, it is not difficult to show that the star-triangle equation (5) also holds. Indeed, given three elements α, β, γ in X, the values of each side of (5) are entirely determined by the mutual equalities between these elements. Moreover, three of the five resulting cases are immediate consequences of equations (1), (2), (4).

Thus when $D^2 = n$, $D = -b^2 - b^{-2}$, the matrices $W^+ = -b^{-3} I + b (J - I)$, $W^- = -b^3 I + b^{-1} (J - I)$ define a symmetric 2-weight spin model. To identify the corresponding link invariant, we observe that $W^+ = b^{-1} DI + bJ$, $W^- = b D I + b^{-1} J$. Consider four unoriented connected link diagrams L^+, L^-, L^0, L^∞ which are everywhere the same except inside a small disk where they behave as shown on Figure 5.

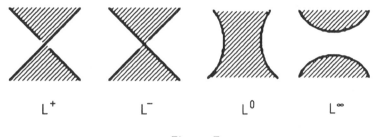

$$L^+ \qquad L^- \qquad L^0 \qquad L^\infty$$

Figure 5

Let e be the edge of $G(L^+)$ corresponding to the crossing appearing in this Figure. Thus in the evaluation of $Z(L^+) = Z(G(L^+),w)$, we must take $w(e) = W^+$. To evaluate $Z(L^0)$ and $Z(L^\infty)$ instead of $Z(L^+)$, we take $w(e) = I$ and $w(e) = J$ without changing any other value of w. Then it is clear from the definiton of a partition function that the equality $W^+ = b^{-1} D I + bJ$ translates into the equality $Z(L^+) = b^{-1} D Z(L^0) + b\ Z(L^\infty)$. Now introducing the normalization factors, we write $<L>$ for $D^{-|V(G(L))|} Z(L)$. Since L^0 has one black region less than L^+ and L^∞, we obtain $<L^+> = b^{-1}<L^0> + b <L^\infty>$. We can show similarly that $<L^-> = b <L^0> + b^{-1}<L^\infty>$. These two equations are interchanged if one exchanges the colors black and white. In other words they are two versions of one single equation.

Since the function L → <L> defined on unoriented diagrams satisfies this equation and is invariant under Reidemeister moves of types II and III, we can identify it with Kauffman's bracket polynomial introduced in [K2] (where Kauffman's variable A is replaced by b^{-1}). As shown in [K2], normalization by the factor $a^{-T(L)} = (-b^3)^{T(L)}$ for oriented link diagrams L then gives the Jones polynomial of [Jo1]. The above description of the Jones polynomial by a Potts model is one aspect of its connection with the Tutte polynomial first exhibited in [Th].

To conclude this section we want to point out the following facts for further reference. The linear span of I and J is a symmetric Bose-Mesner algebra (actually the simplest of these) with a unique duality Ψ given by (19), (20). It is easy to check that $\Psi(W^+) = D\,W^-$, $\Psi(W^-) = D\,W^+$. Finally, the corresponding association scheme has the following property, which we used to simplify the proof of the star-triangle equation : for any three elements α, β, γ in X, the number of elements x which satisfy given scheme relations with α, β, γ only depends on which scheme relations are satisfied by pairs of elements among α, β, γ. We shall call this property *triple regularity*. It can be viewed as a higher order version of the fundamental property (ii) of association schemes.

4.2. Spin models for the two-variable Kauffman polynomial

For more details on the contents of this section the reader is referred to [Ja1], [H].

The two-variable Kauffman polynomial [K3] is an invariant of oriented links which assigns to every oriented link diagram L a Laurent polynomial in two variables a and z of the form $a^{-T(L)}\Lambda(L)$, where $\Lambda(L)$ does not depend on the orientation of L. This invariant is uniquely determined by the following additional property : if L^+, L^-, L^0, L^∞ differ only locally as shown on Figure 5, then

(22) $\Lambda(L^+) - \Lambda(L^-) = z\,(\Lambda(L^0) - \Lambda(L^\infty))$.

(there is actually a second version of the invariant where the minus signs are replaced by plus signs, but it is essentially equivalent to the first one).

We are now interested in spin models whose associated link invariant is an evaluation of the two-variable Kauffman polynomial. It is natural to consider symmetric 2-weight models (X,W^+,W^-) such that the associated partition function $Z(L)$ satisfies the equality $D^{-|V(G(L))|} Z(L) = \Lambda(L)$ for every unoriented link diagram L (Kauffman's variable a will then correspond to the parameter a appearing in equations (1), (2)). Then property (22) is translated (by the same method used for the Jones polynomial in the previous section) into the following equation :

(23) $W^+ - W^- = z (D I - J)$.

Our problem is now to classify the solutions to equations (1)-(5) and (23).

We assume that W^+ is not a linear combination of I and J. Indeed this case has been studied in the previous section. One can observe that the equations $W^+ = b^{-1}D I + bJ$, $W^- = bD I + b^{-1} J$ imply (23) with $z = b^{-1} - b$. This corresponds to the fact that the Jones polynomial is a specialization of the two-variable Kauffman polynomial.

Let **A** be the linear span of $\{I , J, W^+\}$. Thus this set is a basis of **A**. It is clear from (1) - (4) and (23) that **A** is closed under ordinary matrix product and Hadamard product. Hence **A** is a symmetric Bose-Mesner algebra of dimension 3 which contains W^+, W^-. The relations R_1, R_2 of the corresponding scheme form a pair of complementary *strongly regular graphs* on the vertex-set X (see for instance [S]).

Define the linear map Ψ from **A** to itself by the conditions $\Psi(I) = J$, $\Psi(J) = n I$ and $\Psi(W^+) = D W^-$. Then it follows from (23) that

$\Psi(W^-) = D W^- - z (D J - n I) = D (W^- - z (J - D I)) = D W^+$.

This implies that Ψ satisfies (16). It is not difficult to check (working in the basis $\{I, J, W^+\}$) that Ψ also satisfies (17). Thus Ψ is a duality of **A**. We shall say that our strongly regular graphs are *self-dual* . Self-dual strongly regular graphs on n vertices are characterized by the property that the eigenmatrix P satisfies the equation $P^2 = n I$ (for some ordering of the idempotents).

Finally, let us establish the triple regularity property. Extending this property by linearity, we can reformulate it as follows : for any three elements α, β, γ in X, and for any three matrices A, B, C in **A**, the value of $\sum_{x \in X} A[x,\alpha] B[x,\beta] C[x,\gamma]$ only depends on which scheme relations are satisfied by pairs of elements among α, β, γ. Now to prove this, we may choose three bases S_A, S_B, S_C of **A** and assume that the triple (A,B,C) belongs to $S_A \times S_B \times S_C$. We take $S_A = S_B = \{I, J, W^+\}$ and $S_C = \{I, J, W^-\}$.

When one of the matrices A, B, C is I, it is obvious that
$\sum_{x \in X} A[x,\alpha] B[x,\beta] C[x,\gamma]$ has the required property. When one of these matrices is J, the property follows from the fact that **A** is closed under matrix product. The remaining case $A = B = W^+$, $C = W^-$ is given by the star-triangle equation (5).

It is easy to see that triple regularity for graphs is equivalent to the property that for every vertex v, the neighbors of v induce a strongly regular subgraph, and similarly for the vertices non adjacent to v and distinct from it. This property has been thoroughly investigated in [CGS].

We now try to construct a solution to (1)-(5) and (23) from a self-dual strongly regular graph with the triple regularity property. By (1) and (4), we may write $W^+ = aI + bA_1 + cA_2$, $W^- = a^{-1} I + b^{-1}A_1 + c^{-1}A_2$, where A_1, A_2 are the adjacency matrices of the graph and its complement, and b, c are non-zero complex numbers. Equation (23) is then equivalent to the system :

$a - a^{-1} = z (D - 1)$, $b - b^{-1} = c - c^{-1} = - z$.

Since $n \neq 1$ and hence $D \neq 1$, the first equation determines the Kauffman variable z. W^+ should not be a linear combination of I and J and for this we need that $b \neq c$. Thus the equation $x - x^{-1} = - z$ has two distinct solutions: b is one of them and c is the other. To conform with standard notation, we write

(24) $W^+ = a I - t A_1 + t^{-1} A_2$, $W^- = a^{-1} I - t^{-1} A_1 + t A_2$, with $t^2 \neq -1$;

(25) $a - a^{-1} = (t - t^{-1}) (D - 1)$.

Let $\Psi : \mathbf{A} \to \mathbf{A}$ be a linear map satisfying (17). We index the idempotents so that $\Psi(E_i) = A_i$ for $i = 0,1,2$. Assume that $A_1 = kE_0 + sE_1 + rE_2$. It is easy to check that Ψ satisfies (16) if and only if $\Psi(A_1) = nE_1$ and that this reduces to the equations $n = (r-s)^2$, $k = r^2 + r - rs$. The duality Ψ being described in this way, it is not difficult to show that the property $\Psi(W^+) = DW^-$ is equivalent to the system of equations

(26) $a = st + (r+1)t^{-1}$, $D = s - r$.

Once the choice $D = s - r$ is made (note that $n = D^2$ as required), equations (25), (26) reduce to the system

(27) $a = st + (r+1)t^{-1}$, $a^{-1} = st^{-1} + (r+1)t$.

This system has always solutions, provided that for n > 4 our graph is *primitive* (i.e. the graph and its complement are both connected). It is shown in [Ja1] that the triple regularity property implies that any solution to (27) actually defines a 2-weight spin model via (24). What essentially happens is that, as discovered in [CGS], all parameters associated with the triple regularity property are already determined by s, r, and that the values of these parameters are exactly those needed to establish the star-triangle equation (5).

Thus we have obtained a combinatorial description of symmetric 2-weight spin models satisfying (23) in terms of self-dual strongly regular graphs satisfying the triple regularity property. The graphs of this type known so far (excluding complete graphs or their complements and non-primitive graphs on at least 5 vertices) belong to the following list (or to the list of complementary graphs):

(i) Graphs with r + s + 1 = ±1. Then z = 0, $W^+ = W^-$ is a Hadamard matrix and the corresponding value of the Kauffman polynomial is trivial.

(ii) The square. This example is already mentioned in [Jo3]. We obtain an infinite family of spin models describing a simple and well known specialization of the Kauffman polynomial (essentially the generating function for the Tait number of all orientations).

(iii) The pentagon. This example is studied in particular in [Jo2] and the corresponding value of the Kauffman polynomial has a nice topological interpretation (see also the next section 4.3).

(iv) The lattice graphs. The corresponding values of Λ describe the square of Kauffman's bracket polynomial (see section 4.1).

(v) The Higman-Sims graph with s = - 8, r = 2.

It is conjectured in [CGS] that this list is complete, with the possible exception of graphs satisfying $s = - r^2 - 2r$ (or their complements). This last family of graphs is particularly interesting. They can be characterized as the self-dual strongly regular graphs without triangles (this implies triple regularity). Only three instances are known up to complementation: the pentagon (iii), the complement of the Clebsch graph which is of type (i), and the Higman-Sims graph (v). The corresponding parameters a, t in (24) satisfy $a = - t^5$.

As shown in [Ja5], the powerful algebraic theories developed in the quantum group approach to link invariants can be applied to the study of spin models for the Kauffman polynomial and hence to the study of the corresponding strongly regular graphs. We have obtained in particular the following necessary existence condition : either t is a root of unity or $a = \pm t^k$ for some odd integer k. If we could prove that |k| is at most 5, we would solve the conjecture in [CGS].

4.3. Spin models on Abelian groups

Another example due to [GJ] is given in [Jo3]. Let X be a cyclic group of odd size n (with additive notation) and let ω be a primitive n^{th} root of unity. For x, y in X, let $f(x,y) = \omega^{(x-y)^2}$. Easy calculations show that for every α, β, γ in X, $\Sigma_{x \in X} f(\alpha,x) f(x,\beta)^{-1} = n \, \delta(\alpha,\beta)$ and

$$\Sigma_{x \in X} f(\alpha,x) f(x,\beta) f(x,\gamma)^{-1} = K f(\alpha,\beta) f(\alpha,\gamma)^{-1} f(\beta,\gamma)^{-1},$$

where $K = \Sigma_{x \in X} \omega^{x^2}$. Here K is a Gauss sum and $|K|^2 = n$; so $K \neq 0$. Hence if we set $W^+[x,y] = a \, f(x,y)$, $W^-[x,y] = a^{-1} f(x,y)^{-1}$ with $a^2 = D K^{-1}$, equations (1)-(5) hold. So we have a symmetric 2-weight spin model. An interesting topological interpretation of the associated link invariant is established in [GJ]. When n = 3 we have a Potts model which corresponds to a specialization of the Jones polynomial (see section 4.1) and when n = 5 we have a spin model for the Kauffman polynomial which is example (iii) of section 4.2.

Let X be a finite Abelian group written additively. For every i in X define the matrix A_i in M_X by $A_i [x,y] = \delta(i,y-x)$ for every x, y in X. The linear span **A**(X) of the matrices A_i is easily seen to be a Bose-Mesner algebra on X. One can index the characters of X with the elements of X in such a way that, denoting by χ_i the character indexed by i, the equality $\chi_i(j) = \chi_j(i)$ holds for all i, j in X. Let us define the linear map Ψ from **A**(X) to itself by the equalities $\Psi(A_i) = \Sigma_{j \in X} \chi_i(j) A_j$. It is easy to check that Ψ is a duality.

Returning to the spin models of [GJ] described above (so that X is again an odd cyclic group), we see that W^+, W^- belong to **A**(X). Moreover if the characters of X are indexed in such a way that $\chi_i(j) = \omega^{2ij}$ for all i, j in X, the corresponding duality Ψ is such that $\Psi(W^+) = DW^-$.

The next step was taken in [BB2]. This paper gives for every cyclic group X a large family of non-symmetric 2-weight spin models with W^+, W^- in $\mathbf{A}(X)$. These models generalize both the models of [GJ] and the first examples of non-symmetric models obtained in [KMW]. They also have the property that $\Psi(W^+) = DW^-$ for some duality Ψ. A further generalization to the case where X is any Abelian group was obtained in [Ja3]. It is shown there that if t_i, $i \in X$, are non-zero complex numbers satisfying (with the above indexing of characters)
(28) $t_i\, t_j = \chi_i(j)\, t_0\, t_{i+j}$ for every i,j in X
and normalized such that $\sum_{i \in X} t_i^{-1} = D\, t_0$, then
$(X, W^+ = \sum_{i \in X} t_i A_i$, $W^- = \sum_{i \in X} t_i^{-1} A_i^T)$ is a 2-weight spin model. The proof, like the one in [BB2], uses the triple regularity of $\mathbf{A}(X)$. The reader will easily check that $\Psi(W^{-T}) = DW^+$, and hence $\Psi(W^+) = DW^-$.

The equations (28) are solved explicitly in [BBJ] for any Abelian group. In the same paper it is shown that a very original construction of spin models from even rational lattices by Kac and Wakimoto [KW] gives essentially the same spin models as the solutions to (28).

Four-weight spin models with matrices in $\mathbf{A}(X)$ are investigated in [Ja4]. It is shown that when X is a direct product YxY, these spin models are essentially equivalent to other types of well known statistical mechanical models called *vertex models* and *IRF models*.

Other related works are [B3], [WPK], [KMM].

4.4. Nomura's Hadamard spin models

A *Hadamard graph* is a distance-regular graph of diameter 4 with intersection array {4m,4m-1,2m,1; 1,2m,4m-1,4m}. This means that if we define the matrix A_i (i=0,...,4) in M_X (where X is the vertex-set) by setting the entry $A_i[x,y]$ to 1 if the vertices x,y are at distance i and to 0 otherwise, the A_i (i=0,...,4) form the basis of Hadamard idempotents of a Bose-Mesner algebra $\mathbf{A}$ whose parameters can be deduced from the following equations : $A_1^2 = 4m\, A_0 + 2m\, A_2$, $A_1 A_2 = (4m-1)\,(A_1 + A_3)$, $A_1 A_3 = 2m\, A_2 + 4m\, A_4$, $A_1 A_4 = A_3$. There is an essentially bijective correspondence between Hadamard graphs and Hadamard matrices (see for instance [BCN]).

K. Nomura [N1] has associated with every such Hadamard graph some symmetric 2-weight spin models (X, W^+, W^-) which can be defined as follows (for $D = 4\sqrt{m}$) :

(29) $W^+ = \omega \, (- \alpha^3 A_0 + A_1 + \alpha^{-1} A_2 - A_3 - \alpha^3 A_4)$,

where $\omega^4 = 1$ and $\alpha^2 + \alpha^{-2} + 2 \, \omega^2 \sqrt{m} = 0$ (W^- is given by equation (4)). The proof in [N1] of the star-triangle relation establishes and crucially uses the triple regularity of **A**. This Bose-Mesner algebra admits two dualities and each of Nomura's models satisfies $\Psi(W^+) = DW^-$ for one of these dualities Ψ.

It is shown in [Ja3] and [Ja4] that the invariant of links associated with Nomura's models is closely related to the Jones polynomial. More precisely, the value of this invariant for an oriented link L has a simple description in terms of the Jones polynomials of the sublinks of L (a sublink of L is a link obtained from L by deleting some components).

A main ingredient in the proof is the fact that (for fixed ω) the partition function Z(L) depends on the Hadamard graph only via the parameter α appearing in (29), and is given by some rational function of this parameter. This comes from a "matrix-free" approach developed in [Ja3] for the computation of partition functions of plane graphs with edge weights taken in some "exactly triply regular" self-dual Bose-Mesner algebra. The basic idea is as follows. When one computes the partition function of such a spin model on a weighted graph, one can use certain simplification operations. For instance, one can delete a loop if one multiplies at the same time the current value by the (constant) diagonal element of the weight of this loop. Contraction of a pendant edge can be handled similarly using the fact that edge weights have constant row or column sums. One can also replace two parallel edges by a single edge with the same ends weighted with the Hadamard product of the original weights, and matrix multiplication can be used in the same way in relation with the contraction of one edge in a series pair. Changes in edge orientations can be accomodated by corresponding transpositions of edge weights. If our graph is series-parallel, these *series-parallel reductions* allow us to complete the computation of the partition function without enumerating the states. The whole process, and hence its result, only depends on the Bose-Mesner algebra structure (given for instance by the intersection numbers), not on the underlying association scheme.

The triple regularity property allows us to incorporate one more operation : the replacement of a weighted "star" with three ends by a linear combination of weighted triangles. We need another converse operation where a weighted triangle is replaced by a linear combination of weighted stars, and we call *exactly triply regular* the Bose-Mesner algebras where this operation is also available. Then a theorem of Epifanov [E] implies that with the above operations we can reduce every weighted plane graph to a linear combination of trivial such graphs. Thus for exactly triply regular Bose-Mesner algebras one can compute partition functions on all plane weighted graphs without enumerating the states, and the result only depends on the parameters describing the exact triple regularity property. It is shown in [Ja3] that self-dual Bose-Mesner algebras with the triple regularity property are exactly triply regular, and this holds in particular in the case of Hadamard graphs.

To complete the analysis of the partition function for Nomura's models it is then enough to study a special infinite family of examples for which the relation with the Jones polynomial can be established [Ja4].

4.5. Other examples, and some negative results

Other types of 2-weight spin models are discussed in [BB4], and small examples are given. One of these types, called *Hadamard type*, has the property that at least one of the spin model matrices is a Hadamard matrix. In [Y1] infinite families of examples are constructed.

In [I1], [I2], [I3] the possibility of non-symmetric 2-weight spin models in a Bose-Mesner algebra of dimension 3 or 4 is explored. It is shown that this possibility reduces to previously known small spin models.

In [Ja5] we examined the possibility of obtaining 4-weight spin models for the Homfly polynomial of oriented links. The "skein relation" for the Homfly polynomial leads to the consideration of the two new equations $W_1 - W_3 = z\,D\,I$ and $W_4 - W_2^T = z\,J$. Then an interesting connection with symmetric designs appears. Recently this connection was used to show that such models do not exist, except when the Homfly polynomial specializes to the Jones polynomial or to a trivial evaluation [Ja7].

5. TOWARDS A CLASSIFICATION OF SPIN MODELS?

5.1. Two direct approaches

It is possible to obtain some interesting information on spin models by using direct consequences of the invariance equations.

For instance, let us consider the first star-triangle equation in the definition of 4-weight models (see (10)):

for every α, β, γ in X,

$$\sum_{x \in X} W_1[\alpha,x] \ W_1[x,\beta] \ W_4[\gamma,x] = D \ W_1[\alpha,\beta] \ W_4[\gamma,\alpha] \ W_4[\gamma,\beta].$$

Let us fix γ in X, and let Δ be the diagonal matrix in M_X whose (x,x) entry is $W_4[\gamma,x]$. The above equation can then be reformulated as the matrix equality $W_1 \Delta W_1 = D \Delta W_1 \Delta$, or equivalently, since W_1 and Δ are invertible by (8) and (9), $\Delta W_1 \Delta^{-1} = D W_1^{-1}\Delta W_1$. It follows that W_1 and $D\Delta$ are conjugate. Hence every row of DW_4 consists of the spectrum of W_1 in some order. The other equation in (10) gives similarly the same property for the columns. Thus if k denotes the number of distinct eigenvalues of W_1, we can write W_4 as a linear combination with distinct coefficients of k disjoint (0,1) matrices, each of which has constant row and column sums. This combinatorial fact is the main tool used in [BJS] to classify "small" 2-weight spin models (X has up to 7 elements in the symmetric case, 5 otherwise).

Another related approach due to Nomura [N3] consists in "localizing" the star-triangle equation (5) for symmetric 2-weight models in the following way. Let the value θ appear as an entry of W^+. By (4), $D\theta^{-1}$ is an entry of $D W^-$. We have just seen (in a more general setting) that consequently $D\theta^{-1}$ is an eigenvalue of W^+. Moreover, if $m(\theta)$ is the multiplicity of this eigenvalue, $D\theta^{-1}$ appears exactly $m(\theta)$ times in each row and column of DW^-, and consequently θ appears exactly $m(\theta)$ times in each row and column of W^+. Now let N be the matrix with rows indexed by X and columns indexed by XxX whose $(\alpha,(\beta,\gamma))$ entry is $W^+[\alpha,\beta] \ W^-[\alpha,\gamma]$. By (5), the (β,γ)-column of N is an eigenvector of W^+ for the eigenvalue $DW^-[\beta,\gamma]$. Hence the rank of the submatrix of N corresponding to all columns (β,γ) such that $W^+[\beta,\gamma] = \theta$ is at most $m(\theta)$.

Note that this submatrix has n rows and n m(θ) columns. Thus many determinants of size m(θ)+1 must vanish, yielding useful equations for the entries of W$^+$. This method is used in [N3] to classify spin models for which some multiplicity m(θ) is 1 or 2 (with a few additional assumptions) and in [N5] (with a slight extension) to show that if a symmetric 2-weight spin model is defined in a natural way from the distance function on the vertices of a triangle-free graph, this graph is distance-regular.

5.2. Association schemes : a natural framework for spin models

We have observed in section 4 that the known two-weight spin models have the following properties : there is some Bose-Mesner algebra **A** with a duality Ψ such that W$^+$, W$^-$ belong to **A**, and Ψ(W$^+$) = D W$^-$. It was already shown in [Ja1] that this situation is very natural. To every symmetric 2-weight spin model (X,W$^+$, W$^-$) corresponds a pair **M**, **H** of algebras which are vector subspaces of M$_X$: **M** is generated by W$^+$ and J with product the ordinary matrix product, and **H** is generated by W$^-$ and I with product the Hadamard product. Then there exists a unique isomorphism of algebras Ψ from **M** to **H** such that Ψ(W$^+$) = D W$^-$ and Ψ(J) = n I. Moreover Ψ is given by the following expression :

(30) For every M in **M**, Ψ(M) = D^{-1}a W$^+$ (W$^-$o (W$^+$ M))

When **M** and **H** are equal, **M** is the required Bose-Mesner algebra and Ψ is the required duality. The reader for instance will be able to check easily formula (30) for the examples in sections 4.1 and 4.2. The above results were soon generalized in [BB1] to non-symmetric 2-weight models and 4-weight models.

We have now simpler proofs of these results for 2-weight models [BBJ]. We restrict our attention to the symmetric case. It is not difficult to show that the star-triangle equation (5) can be reformulated as follows :

(31) For every M in M$_X$, W$^+$ (W$^-$o (W$^+$ M)) = DW$^-$ o (W$^+$(W$^-$ oM)).

Now we define the linear map Ψ : M$_X \to$ M$_X$ by the formula (30).

Then one can show directly that Ψ defines an algebra isomorphism from **M** to **H** such that Ψ(W$^+$) = D W$^-$ and Ψ(J) = n I. The last equality is immediate from equations (2), (3). To indicate how the proof of the isomorphism property works, let us show that Ψ(W$^-$M) = Ψ(W$^-$)oΨ(M).

By (30) and (3), $\Psi(W^-M) = Da\ W^+ (W^- o\ M)$ and, when $M = I$, $\Psi(W^-) = D\ W^+$. On the other hand, by (30) and (31), $\Psi(M) = aW^- o\ (W^+(W^- oM))$ and the result now follows from (4). The second formula for Ψ easily gives that $\Psi(W^+) = D\ W^-$.

One shows similarly that $n^{-1}\Psi$ is an algebra isomorphism from **H** to **M**, and that when $M = H = A$, Ψ defines a duality on **A**.

This already motivates some special interest for spin models whose weight matrices belong to some Bose-Mesner algebra. Moreover in this situation we have a great simplification of equations (1)-(4) (and, if valid, the triple regularity property also greatly simplifies the star-triangle equation (5)). A good starting point seems to be to consider simpler special cases. For instance in [I3] the classification by [No] of association schemes of size at most 10 is used to classify all corresponding 2-weight spin models with size at most 8. Another example is the classification in [N4] of the symmetric 2-weight spin models for which **M** = **H** is the Bose-Mesner algebra of some bipartite distance-regular graph .

5.3. Modular invariance

We also want to take into account the duality property. This will involve the concept of *modular invariance* for self-dual Bose-Mesner algebras. This concept was discovered by Eiichi Bannai and his coworkers [B1] when they established an equivalence between the abstract versions of the following two kinds of objects : fusion algebras of conformal field theories, and Bose-Mesner algebras as discussed here. Then the modular invariance property for Bose-Mesner algebras corresponds via this equivalence to the crucial modular invariance property for fusion algebras. When the Bose-Mesner algebra is self-dual with a duality described by the eigenmatrix P, the modular invariance property can be stated as follows :

(32) there exists a diagonal matrix T such that $(PT)^3 = I$.

Then investigations by the same research group showed that solutions to the modular invariance equations could give spin models. The first example was that of Hamming schemes ([BB3], [BBIK]), and the second was that of cyclic group schemes [BB2].

The relationship between 2-weight spin models, the previous duality results of [Ja1], [BB1] and the modular invariance property was finally understood as follows (see [BBJ]). Again we consider only the symmetric case for the sake of simplicity. Thus let (X, W^+, W^-) be a symmetric 2-weight spin model. We assume that $\mathbf{M} = \mathbf{H} = \mathbf{A}$ and so, as we have just seen, $\mathbf{A}$ has a duality Ψ given by $\Psi(M) = a\, W^- \circ (W^+(W^- \circ M))$. Let P be the eigenmatrix of $\mathbf{A}$ corresponding to the duality Ψ ; that is, P is the matrix of Ψ with respect to the basis $\{E_i , i = 0,...d\}$ of orthogonal idempotents for the ordinary matrix product (see section 3). Thus $\Psi(E_i) = A_i$ and $\Psi(A_i) = nE_i$. We can also view P as the transition matrix from $\{A_i , i = 0,...d\}$ (the basis of Hadamard idempotents) to $\{E_i , i = 0,...d\}$. Now write $W^- = \sum_{i=0,...,d} t_i A_i$. The matrix of the linear map $M \to W^- \circ M$ on $\mathbf{A}$ with respect to the basis $\{A_i, i = 0,...d\}$ is the diagonal matrix T with $T(i,i) = t_i$ for $i = 0,...d$. Note that $W^+ = D^{-1}\Psi(W^-) = D \sum_{i=0,...,d} t_i E_i$. Hence the matrix of the linear map $M \to W^+ M$ on $\mathbf{A}$ with respect to the basis $\{E_i, i = 0,...d\}$ is DT, and the matrix of this map with respect to the basis $\{A_i, i = 0,...d\}$ is $P^{-1}(DT)P = D^{-1}PTP$. Thus the matrix of Ψ with respect to the basis $\{A_i, i = 0,...d\}$ is $D^{-1}aTPTPT$. This matrix is also $P^{-1}(P)P = P$. This yields the equation $(PT)^3 = D^3 a^{-1}I$.

Thus the solutions T to the modular invariance equations (32) correspond exactly (up to normalization) to the pairs of matrices W^+, W^- such that $\Psi(M) = a\, W^- \circ (W^+(W^- \circ M))$ for all M in $\mathbf{A}$. It is shown in [BBJ] that these pairs satisfy equations (1)-(4) and also the matrix version (31) of the star-triangle equation, but restricted to matrices M in $\mathbf{A}$. When $\mathbf{A}$ is $\mathbf{A}(X)$ for some Abelian group X, this restriction is not significant and we obtain 2-weight spin models. They are exactly those described by equation (28) in section 4.3.

So clearly a good approach to the construction of 2-weight spin models is to investigate solutions of the modular invariance equation for self-dual Bose-Mesner algebras and study the star-triangle equation for the corresponding matrices. How general is this approach? The following answer for the symmetric case was found quite recently.

<u>Theorem</u> : Let (X, W$^+$, W$^-$) be a symmetric 2-weight spin model.
Then there exists a symmetric Bose-Mesner algebra **A** on X which contains W$^+$, W$^-$. Moreover the map Ψ from **A** to itself defined by Ψ(M) = a W$^-$o (W$^+$(W$^-$o M)) is a duality on **A**. Thus the spin model corresponds to a solution of the modular invariance equations for some duality in **A**.

The proof in [Ja6] has a strong topological flavor : the algebra **A** is obtained as the image of a certain algebra of "tangles" under a matrix-valued map which extends the concept of a partition function. However one can then consider the smallest symmetric Bose-Mesner algebra containing W$^+$, W$^-$, which also has the properties stated in the Theorem. This "smaller" Bose-Mesner algebra can be described explicitly and computed efficiently. Immediately afterwards, K. Nomura found a purely algebraic proof of the existence of a symmetric Bose-Mesner algebra containing W$^+$, W$^-$ [N6], and we have every reason to believe that the duality and modular invariance properties can be proved in the same way. Nomura's definition of his Bose-Mesner algebra is strikingly simple : it is the set of symmetric matrices for which every column of the matrix N introduced in section 5.1 is an eigenvector. At the time of this writing we do not know whether the three above-mentioned Bose-Mesner algebras can actually be different. The above theorem lays a firm basis for the classification of symmetric spin models.

The modular invariance equations are solved for the Bose-Mesner algebras of the examples in section 4. The same is true for the symmetric subalgebra of **A**(X) when X is a cyclic group [B3], the Bose-Mesner algebras of Hamming schemes [BB3], and the Bose-Mesner algebras of 2-Sylow subgroups of Suzuki simple groups Sz(q) [BBJ]. Recently the case of self-dual distance-regular graphs was thoroughly studied in [CS]. This paper shows that for these Bose-Mesner algebras the modular invariance equation has at most 12 solutions, and that there are no solutions (except for small sizes) for the graphs of bilinear, alternating and Hermitian forms.

It is possible that other properties of self-dual Bose-Mesner algebras can also be used as necessary conditions for the existence of spin models. The occurence of the triple regularity property in the examples of section 4 cannot be generalized, as shown by the Hamming scheme spin models. Another property called *planar duality* is introduced in [Ja3] (see also [Bi1]) and holds for the Bose-Mesner algebra **A** constructed in [Ja6] by the tangle approach. Roughly speaking, this property means that the partition function of a spin model with edge weights in **A** on a connected plane graph is equal (up to a simple factor) to the partition function on the dual plane graph, where edge weights are "dualized" using the duality of **A**. The case when the plane graph is the 4-clique is of special significance.

5. 4. Constructions

A sound classification of spin models should take into account the existing operations on the class of these objects. For instance there is a natural tensor product operation (see [H], section 5.3) which corresponds to the product of partition functions, and the spin models which can be found in the Bose-Mesner algebras of Hamming schemes turn out to decompose into products of Potts models [BBIK]. More mysterious "twisted" versions of the tensor product of a spin model with a spin model of size 4 are introduced in [N2]. A construction in [Ja4] inspired by relations with vertex and IRF models is associated with a certain composition of link invariants and gives as a special case the interpretation of the partition function of Nomura's models of section 4.4. Some constructions given in [N2] and [Y2] also relate certain types of 4-weight models with certain types of 2-weight models. Finally some transformations which preserve the partition function (inspired by the gauge transformations in physics) should be considered as well. For instance in [Ja8] such transformations are used to show that the non-symmetric 2-weight spin models of section 4.3 actually give the same link invariants as the symmetric ones.

REFERENCES

[B1] Ei. BANNAI, Association schemes and fusion algebras (an introduction), J. of Algebraic Combinatorics 2 (1993), 327-344.

[B2] Ei. BANNAI, Algebraic Combinatorics-recent topics on association schemes, Sugaku 45 (1993), 55-75 (in Japanese). English translation to appear in Sugaku Expositions, AMS, 1995.

[B3] Et. BANNAI, Modular invariance property and spin models attached to cyclic groups association schemes, J. of Stat. Planning and Inference, to appear.

[BB1] E. BANNAI, E. BANNAI, Generalized spin models and association schemes, Mem. Fac. Science Kyushu Univ. A, 47, n° 2 (1993), 397-409.

[BB2] E. BANNAI, E. BANNAI, Spin models on finite cyclic groups, J. of Algebraic Combinatorics 3 (1994), 243-259.

[BB3] E. BANNAI, E. BANNAI, Modular invariance of the character table of the Hamming association scheme H(d,q), J. Number Theory 47 (1994), 79-92.

[BB4] E. BANNAI, E. BANNAI, Generalized generalized spin models (four-weight spin models), Pacific J. of Math., to appear.

[BBIK] E. BANNAI, E. BANNAI, T. IKUTA, K. KAWAGOE, Spin models constructed from the Hamming association schemes, Proceedings 10th Algebraic Combinatorics Symposium, Gifu Univ. (1992), 91-106.

[BBJ] E. BANNAI, E. BANNAI, F. JAEGER, On spin models, modular invariance, and duality, submitted.

[BI] Ei. BANNAI, T. ITO, Algebraic Combinatorics I, Association schemes, Benjamin/Cummings, Menlo Park, 1984.

[BJS] Ei. BANNAI, F. JAEGER, A. SALI, Classification of small spin models, Kyushu J. Math. 48 (1994), 185-200.

[Ba] R.J. BAXTER, Exactly solved models in statistical mechanics, Academic Press, 1982.

[Bi1] N.L. BIGGS, On the duality of interaction models, Math. Proc. Cambridge Philos. Soc. 80 (1976), 429-436.

[Bi2] N.L. BIGGS, Interaction models, London Math. Soc. Lecture Notes 30, Cambridge University Press, 1977.

[BM] R.C. BOSE, D.M. MESNER, On linear associative algebras corresponding to association schemes of partially balanced designs, Ann. Math. Statist. 30 (1959), 21-38.

[BCN] A.E. BROUWER, A.M. COHEN, A. NEUMAIER, Distance-regular graphs, Springer-Verlag, Ergebnisse der Mathematik und ihrer Grenzgebiete, 3. Folge, Band 18, 1989.

[BZ] G. BURDE, H. ZIESCHANG, Knots, de Gruyter, Berlin, New York, 1985.

[CGS] P. J. CAMERON, J. M. GOETHALS, J.J. SEIDEL, Strongly regular graphs having strongly regular subconstituents, J. Algebra 55 (1978), 257-280.

[CS] L. CHIHARA, D. STANTON, A matrix equation for association schemes, Graphs and Combinatorics, to appear.

[D] P. DELSARTE, An algebraic approach to the association schemes of coding theory, Philips Research Reports Supplements 10 (1973).

[E] G.V. EPIFANOV, Reduction of a plane graph to an edge by a star-triangle transformation, Soviet Math. Doklady 7 (1966), 13-17.

[GJ] D.M. GOLDSCHMIDT, V.F.R. JONES, Metaplectic link invariants, Geom. Dedicata 31 (1989), 165-191.

[H] P. de la HARPE, Spin models for link polynomials, strongly regular graphs and Jaeger's Higman-Sims model, Pacific J. of Math. 162 n°1 (1994), 57-96.

[HJa] P. de la HARPE, F. JAEGER, Chromatic invariants for finite graphs : theme and polynomial variations, submitted.

[HJo] P. de la HARPE, V.F.R. JONES, Graph invariants related to statistical mechanical models : examples and problems, J. Comb.Theory Ser. B 57 (1993) n°2, 207-227.

[I1] T. IKUTA, Spin models corresponding to nonsymmetric association schemes of class 2 , Mem. Fac. Sci. Kyushu Univ. Ser A, 47, n° 2 (1993), 383-390.

[I2] T. IKUTA, Spin models corresponding to nonsymmetric association schemes of class three, preprint, 1993.

[I3] T. IKUTA, On spin models attached to association schemes, PhD thesis, Kyushu University, 1994.

[Ja1] F. JAEGER, Strongly regular graphs and spin models for the Kauffman polynomial, Geom. Dedicata 44 (1992), 23 -52.

[Ja2] F. JAEGER, Modeles a spins, invariants d'entrelacs, et schemas d'association, Actes du Seminaire Lotharingien de Combinatoire, 30^{ieme} session, R. Konig et V. Strehl ed., IRMA, Strasbourg, 1993, 43-60.

[Ja3] F. JAEGER, On spin models, triply regular association schemes, and duality, J. of Algebraic Combinatorics, to appear.

[Ja4] F. JAEGER, New constructions of models for link invariants, Pacific J. of Math., to appear.

[Ja5] F. JAEGER, Spin models for link invariants and representations of algebras, Int. Conference on Algebraic Combinatorics, Fukuoka, 1993.

[Ja6] F. JAEGER, Classification of spin models in terms of association schemes, in preparation.

[Ja7] F. JAEGER, On 4-weight spin models for the Homfly polynomial, in preparation.

[Ja8] F. JAEGER, Transformations of spin models which preserve the partition function, in preparation.

[Jo1] V.F.R. JONES, A polynomial invariant for knots via Von Neumann algebras, Bull. Am. Math. Soc. 12 (1985), 103-111.

[Jo2] V.F.R. JONES, On a certain value of the Kauffman polynomial, Comm. Math. Phys. 125 (1989), 459-467.

[Jo3] V.F.R. JONES, On knot invariants related to some statistical mechanical models, Pacific J. of Math. 137 (1989) 2, 311-334.

[KW] V.G. KAC, M. WAKIMOTO, A construction of generalized spin models, preprint, 1993.

[K1] L.H. KAUFFMAN, On Knots, Annals of Mathematical Studies 115, Princeton University Press, Princeton, New Jersey, 1987.

[K2] L.H. KAUFFMAN, State models and the Jones polynomial, Topology 26 (1987), 395-407.

[K3] L.H. KAUFFMAN, An invariant of regular isotopy, Trans. AMS 318 (2) 1990, 417-471.

[KMW] K. KAWAGOE, A. MUNEMASA, Y. WATATANI, Generalized spin models, J. of Knot Theory and its Ramifications, 3 (1994) n°4, 465-475.

[KMM] T. KOBAYASHI, H. MURAKAMI, J. MURAKAMI, Cyclotomic invariants for links, Proc. Japan Acad. 64 A (1988), 235-238.

[No] E. NOMIYAMA, Classification of association schemes with at most ten vertices, preprint; Master's degree thesis, Kyushu Univ.,1994.

[N1] K. NOMURA, Spin models constructed from Hadamard matrices, J. Comb. Theory Ser. A, 68 (1994), 251-261.

[N2] K. NOMURA, Twisted extensions of spin models, J. of Algebraic Combinatorics, to appear.

[N3] K. NOMURA, Spin models with an eigenvalue of small multiplicity, J. Comb. Theory Ser. A, to appear.

[N4] K. NOMURA, Spin models on bipartite distance regular graphs, submitted, 1994.

[N5] K. NOMURA, Spin models on triangle-free connected graphs, submitted, 1994.

[N6] K. NOMURA, An algebra associated with a spin model, preprint, 1994.

[S] J.J. SEIDEL, Strongly regular graphs, in : Surveys in Combinatorics, B. Bollobas editor, pp.157-180, LMS Lecture Notes Series 38, Cambridge University Press, 1979.

[T] H.N.V. TEMPERLEY, Lattice models in discrete statistical mechanics, in "Applications of Graph Theory", R.J. Wilson and L.W. Beineke editors, pp. 149-175, Academic Press, London, 1979.

[Th] M.B. THISTLETHWAITE, A spanning tree expansion of the Jones polynomial, Topology 26 (1987), 297-309.

[Tur] V.G. TURAEV, The Yang-Baxter equation and invariants of links, Invent. Math. 92 (1988), 527-553.

[Tut] W.T. TUTTE, On dichromatic polynomials, J. Comb. Theory 2 (1967), 301 - 320.

[WPK] F.Y.WU, P. PANT, C.KING, A new link invariant from the chiral Potts model, preprint, 1994.

[Y1] M. YAMADA, Hadamard matrices and spin models, J. of Stat. Planning and Inference, to appear.

[Y2] M. YAMADA, The construction of 4-weight spin models by using Hadamard matrices and M-structure, Australasian J. of Comb., to appear.

Computational Pólya theory

Mark Jerrum[*]
Department of Computer Science
University of Edinburgh
United Kingdom

Abstract

A permutation group G of degree n has a natural induced action on words of length n over a finite alphabet Σ, in which the image x^g of x under permutation $g \in G$ is obtained by permuting the positions of symbols in x according to g. The key result in "Pólya theory" is that the number of orbits of this action is given by an evaluation of the *cycle-index polynomial* $P_G(z_1, \ldots, z_n)$ of G at the point $z_1 = \cdots = z_n = |\Sigma|$. In many cases it is possible to count the number of essentially distinct instances of a combinatorial structure of a given size by evaluating the cycle-index polynomial of an appropriate symmetry group G.

We address the question "to what extent can Pólya theory be mechanised?" There are compelling complexity-theoretic reasons for believing that there is no efficient, uniform procedure for computing the cycle-index polynomial exactly, but less is known about approximate evaluation, say to within a specified relative error. The known results — positive and negative — will be surveyed.

1 Preliminaries

This article is concerned with a topic in computational algebra, which combines aspects of combinatorics, algorithmics, and computational complexity. On the assumption that most readers will be unfamiliar with at least one of these, the first section aims to give a brief account of key facts.

---

[*]The author is supported in part by grant GR/F 90363 of the UK Science and Engineering Research Council, and by Esprit Working Group No. 7097, "RAND." Address for correspondence: Department of Computer Science, University of Edinburgh, The King's Buildings, Edinburgh EH9 3JZ, United Kingdom.

1.1   Elementary group-theoretic preliminaries

Let Σ be a finite alphabet of cardinality k, and G a permutation group on $[n] = \{0,\ldots,n-1\}$. The group G has a natural induced action on the set Σ^n of all words of length n over the alphabet Σ, under which the word $\alpha = x_0 x_1 \ldots x_{n-1}$ is mapped by the permutation $g \in G$ to the word $\alpha^g = y_0 y_1 \ldots y_{n-1}$ defined by $y_j = x_i$ for all $i, j \in [n]$ satisfying $i^g = j$.[1] The action of G partitions Σ^n into a number of *orbits*, these being the equivalence classes of Σ^n under the equivalence relation that identifies α and β whenever there exists $g \in G$ mapping α to β.

By way of example, let $\Sigma = \{0,1\}$, and $n = m^2$ for some integer m. Interpret each element of Σ^n as the adjacency matrix of an m-vertex directed graph Γ. Let G be the permutation group of degree n and order $m!$ whose elements correspond to permutations of the vertex set of Γ, i.e., to simultaneous permutations of the rows and columns of the adjacency matrix. Then the orbits of Σ^n under the action of G correspond naturally to unlabelled directed graphs on m vertices. Many other sets of unlabelled combinatorial structures can be obtained using a like construction.

What has just been described is the setting for *Pólya theory*, the key result of which is an expression for the number of orbits in terms of the *cycle-index polynomial* of G [4]. This polynomial, in the variables $z_1, z_2, \ldots, z_n$, is defined to be

$$P_G(z_1,\ldots,z_n) = |G|^{-1} \sum_{g \in G} z_1{}^{c_1(g)} \ldots z_n{}^{c_n(g)}, \qquad (1)$$

where $c_i(g)$ denotes the number of cycles in g of length i. The key result just referred to is that the number of orbits of Σ^n under the action of G is P_G evaluated at the point $z_1 = \cdots = z_n = k$. For many important choices for G, this computation is feasible and leads to results concerning the number of unlabelled combinatorial structures of various kinds [11]. In this article we study the problem of computing P_G for an arbitrary permutation group G and at an arbitrary point.

1.2   Algorithmic preliminaries

It is clear that the cycle-index polynomial of a permutation group G may be computed at a specified point $z_1 = a_1, \ldots, z_n = a_n$, directly from definition (1) by explicit enumeration of the elements of G. In general, however, the order of G is exponentially larger than its description, say, as a list of generators. The interesting question from a computational viewpoint is whether some efficient algorithm exists for computing the cycle-index polynomial that avoids

---

[1] Note that the permutation g is considered to act on *positions* rather than *indices*, since this is perhaps the easier to grasp of the two possible conventions.

explicit enumeration. In the theoretical computer science tradition, we take as a first approximation to the notion of "efficient algorithm" a procedure (expressed as a Turing machine, or as a program in some simple programming language) that runs in time polynomial in some natural measure of input size.

In this article we take as our measure of input size simply the degree n of the permutation group G presented as input. This choice requires justification. On the one hand, a simple counting argument demonstrates that $\Omega(n \log n)$ bits are required to encode all permutation groups of degree n, so it is unreasonable to choose a measure of input size substantially smaller than n. On the other hand, as we shall see presently, every permutation group G of degree n has a simple and compact (in terms of n) encoding that allows many questions about G to answered efficiently.

Following Sims [19], for $1 \leq i \leq n$ let

$$G_i = \{g \in G : j^g = j, \text{ for all } 0 \leq j < i\},$$

be the subgroup of G stabilising $[i]$ pointwise, and let $G_0 = G$. Then

$$G = G_0 \geq G_1 \geq \cdots \geq G_n = \{1\}$$

is a sequence of subgroups of G with the property that the index $|G_i : G_{i+1}|$ of G_{i+1} in G_i is bounded by n. For $0 \leq i \leq n-1$, let U_i be a right transversal for $G_i : G_{i+1}$, i.e., a set that contains precisely one element from each right coset of G_{i+1} in G_i. The collection $\{U_i\}$ is called a *strong generating set* for G. Note that a strong generating set contains only $O(n^2)$ permutations in total.

Aside from economy of space, a strong generating set has the advantage of efficiently supporting various operations, of which the most basic is membership testing. To decide whether permutation g is member of G, search for a permutation $u \in U_0$ such that gu^{-1} stabilises 0. If no such permutation exists, then $g \notin G$; otherwise the permutation u is unique, and $g \in G$ iff $gu^{-1} \in G_1$. Now recursively test whether $gu^{-1} \in G_1$ using the observation that $\{U_i : 1 \leq i \leq n-1\}$ is a strong generating set for G_1. Making realistic assumptions about the model of computation, the decision procedure just sketched can be implemented to run in time $O(n^2)$.

Furst, Hopcroft, and Luks [7], who were the first to analyse Sims' data structure from a complexity-theoretic viewpoint, showed that a strong generating set may be computed in time $O(n^6)$ from an arbitrary (small) set of generators for G. This bound on time-complexity was subsequently improved to $O(n^5)$ by the author [12] using elementary techniques, and to $\tilde{O}(n^4)$ by Babai, Luks, and Seress [2], using non-elementary techniques relying on the classification of finite simple groups.[2]

---

[2]The $\tilde{O}()$ notation hides not merely constants, but also arbitrary powers of $\log n$.

1.3   Complexity-theoretic preliminaries

Let Δ be a finite alphabet, perhaps the binary alphabet, in which we agree to encode problem instances, whether they be groups, graphs, numbers, or whatever. We are concerned with computational problems whose solution is a natural number; abstractly, such problems may be viewed as functions from instances Δ^* to solutions N. In fact, for reasons that will become apparent shortly, it will prove convenient to consider a slightly more general set-up in which the instance consists of a number m of recognisable parts, all encoded over the alphabet Δ.

We say that a function $f : (\Delta^*)^m \to \mathsf{N}$ is *polynomial time* if there exists a Turing machine that computes $f(\alpha_1, \ldots, \alpha_m)$ in time polynomial in $|\alpha_1| + \cdots + |\alpha_m|$. The complexity class #P was introduced by Valiant [22, 23] as a counting analogue to the more familiar class NP of decision problems. It may be defined in several equivalent ways, of which the following is particularly well suited to our purpose. A function $f : \Delta^* \to \mathsf{N}$ is contained in #P precisely if it can be expressed in the form

$$f(x) = \sum_{y \in \Delta^{p(|x|)}} w(x, y), \tag{2}$$

where p is a univariate polynomial, and $w : (\Delta^*)^2 \to \mathsf{N}$ is a polynomial-time "weight function."

For example, we might interpret the word x as the encoding of an undirected graph G, the word y as the encoding of a subgraph of G, and define $w(x, y)$ to be 1 if y encodes a perfect matching in G, and 0 otherwise. Then the function f defined by (2) counts the number of perfect matchings in a graph. Since the function w is clearly polynomial time, we may deduce that the problem of counting perfect matchings in a graph is in the class #P. As the specific combinatorial structure "perfect matching" in the above example could be replaced by almost any other (positively weighted) structure, it will be apparent that #P is a wide ranging class, which contains many problems of combinatorial interest.

Just as with NP, the class #P contains "complete" functions that are efficiently universal for the class, and hence (informally) as hard to compute as any function in the class. A function f' is *polynomial-time Turing reducible to f* if there is a polynomial-time Turing machine that computes f' given an oracle for f.[3] A function f is *#P-hard* if every function in #P is Turing reducible to f; it is *#P-complete* if, in addition, $f \in$ #P. The key observation is that if *any* #P-hard function is computable in polynomial time, then *every* function in #P is computable in polynomial time. Since #P is a wide-ranging

---

[3] An *oracle* for a function f is a black box that takes as input a word $x \in \Delta^*$ and in one time-step produces as output $f(x)$.

class containing many apparently hard-to-compute functions, #P-hardness (or completeness) may be interpreted as strong evidence for intractability.

Valiant [22] showed that counting perfect matchings in a bipartite graph is #P-complete and hence likely to be computationally intractable, even though there is a polynomial-time procedure for deciding existence of a perfect matching, using the classical "augmenting path" technique. This was the first really convincing demonstration of a phenomenon that is now recognised as widespread: that a combinatorial counting problem may be computationally intractable, even when the allied decision problem is computationally easy. The catalogue of known #P-complete problems is now very extensive.

2 The complexity of counting modulo a group of symmetries

For an infinite sequence $(a_i : i \in \mathbb{N}^+)$ of rational numbers, let CYCLEINDEX(a_i) be the problem:

INSTANCE (an encoding of) a permutation group G of degree n;

OUTPUT the value $|G| P_G(a_1, \ldots, a_n)$.

The factor $|G|$ in the output makes the problem into a straight generating function evaluation. Its addition is a matter of convenience, and has no essential effect on the computational difficulty of the problem, as the order $|G|$ of a group G is easy to compute from a strong generating set for G. It is straightforward to verify, directly from the definition of the class #P, that CYCLEINDEX$(k, k, \ldots) \in$ #P, for any positive integer k.

Three cases are known in which CYCLEINDEX(a_i) is easy to compute:

(i) if $a_i = 0$ for all $i > 1$, then only the identity permutation carries any weight and the required output is a_1^n;

(ii) if (a_i) is a geometric progression with initial element r and common ratio r, then every permutation carries equal weight, and the required output is $|G| r^n$;

(iii) if (a_i) is a geometric progression with initial element r and common ratio $-r$, then all even (respectively, odd) permutations carry weight r^n (respectively, $-r^n$), and the required output is either 0 or $|G| r^n$ depending on whether or not G contains odd permutations.

Note that all three cases are algorithmically trivial. Goldberg [9, Thm 22] has shown that for "almost all" non-negative sequences (a_i) not covered by the above cases (for details see Theorem 2), CYCLEINDEX(a_i) is #P-hard, and hence likely to be computationally intractable. The proof of Theorem 2 is technically intricate, so we content ourselves here with proving a special case of particular interest, namely ($a_i = 2$). In doing so, we hope to illustrate the concepts introduced in Section 1.3, and gain insight into why the cycle-index polynomial is hard to compute.

Theorem 1 CYCLEINDEX($2, 2, \ldots$) *is #P-complete.*

Proof   The following problem, which we refer to as #INDSET, was included in Valiant's original list [23] of #P-complete problems, in the guise of "Monotone 2-Sat:"

INSTANCE an undirected graph Γ;

OUTPUT the number of independent sets (not necessarily maximal) in Γ.

Our proof strategy is to exhibit a polynomial-time Turing reduction from #INDSET to CYCLEINDEX($2, 2, \ldots$). Since the relation "is polynomial-time Turing reducible to" is transitive, it will follow that CYCLEINDEX($2, 2, \ldots$) is #P-complete.

Let $\Gamma = (V, E)$ be an undirected graph with vertex set V and edge set E, viewed as an instance of #INDSET, and let $n = |V|$ and $m = |E|$. Arbitrarily orient the edges of G so that each edge $e \in E$ has a defined start vertex e^- and end vertex e^+. Construct a permutation group $G = G(\Gamma)$ of degree $4m$ on base set

$$\Omega = \Big\{ \langle e, 00 \rangle, \langle e, 01 \rangle, \langle e, 10 \rangle, \langle e, 11 \rangle : e \in E \Big\}$$

as follows. For $e \in E$, define (in cycle notation)

$$h_e = (\langle e, 00 \rangle, \langle e, 01 \rangle)\, (\langle e, 10 \rangle, \langle e, 11 \rangle)$$

and

$$h'_e = (\langle e, 00 \rangle, \langle e, 10 \rangle)\, (\langle e, 01 \rangle, \langle e, 11 \rangle).$$

Then $G = \langle g_v : v \in V \rangle$, where, for each $v \in V$, the generator g_v is given by

$$g_v = \prod_{e:e^- = v} h_e \prod_{e:e^+ = v} h'_e.$$

Observe that the generators g_v commute and have order 2. It follows that every group element $g \in G$ may be expressed in the form $g = g_U = \prod_{u \in U} g_u$, for some $U \subseteq V$. It is easy to see that the set U is uniquely determined by g, and hence that $|G| = 2^n$.

For any $U \subseteq V$, the cycle structure of g_U is $1^{4i} \cdot 2^{2(m-i)}$ where i is the number of edges in Γ with both endpoints in the complement of U. Denote by N_i the number of subsets $U \subseteq V$ such that precisely i edges of Γ have both endpoints in the complement of U. Note that N_0 is the number of vertex covers in Γ (and hence, by complementation, of independent sets). Then, for j a positive integer,

$$|G|\, P_G(2^j, 2^j, \ldots, 2^j) = \sum_{i=0}^{m} N_i\, 2^{2(m+i)j} = 2^{2mj} p(2^{2j}),$$

where p is the polynomial $p(\xi) = \sum_{i=0}^{m} N_i \xi^i$. If the values $|G|\, P_G(2^j, \ldots, 2^j)$ for $j = 1, \ldots, m+1$ were known, it would be possible to interpolate to determine $p(0) = N_0$. But

$$P_G(2^j, 2^j, \ldots, 2^j) = P_{G \wr \{1_j\}}(2, 2, \ldots, 2),$$

where $G \wr \{1_j\}$ denotes the wreath (Kranz) product of G with the trivial permutation group on j letters.

Thus, by making $m + 1$ calls to an oracle for CYCLEINDEX$(2, 2, \ldots)$ and performing an interpolation of a univariate polynomial of degree m, it is possible to evaluate the number of independent sets in the graph Γ. The procedure just described constitutes a polynomial-time Turing reduction from #INDSET to CYCLEINDEX$(2, 2, \ldots)$.   □

As indicated earlier, Theorem 1 is a special case of a general result of Goldberg.

Theorem 2 *Let (a_i) be a sequence of non-negative rational numbers satisfying $a_i \neq a_1^i, 0$, for some i. Then* CYCLEINDEX(a_i) *is #P-hard.*

A proof may be found in [9, p. 150] or [10]. Note that we cannot claim #P-completeness here; however, if we assume in addition that (a_i) is a sequence of natural numbers computable in time polynomial in i, then CYCLEINDEX(a_i) is in #P, and hence #P-complete.

It is reasonable to conjecture that the problem CYCLEINDEX(a_i) is #P-hard for all sequences not covered by cases (i)–(iii) identified at the beginning of the section, but this is not known for certain. There is unlikely to be a major conceptual difficulty in resolving the issue, but the solution may require a case analysis of some technical complexity. Note that Theorem 2 already covers all non-trivial non-negative sequences, except those obtained from a geometric progression by setting an arbitrary set of elements to 0.

3   Approximate counting and uniform sampling

We have seen that in almost all its non-trivial variants, the problem of evaluating the cycle-index polynomial P_G *exactly* is #P-hard. This should perhaps not surprise us, since the problem of determining whether a permutation group contains a permutation with a specified cycle structure is known to be NP-complete. However, we might reasonably ask whether there is an efficient *approximation* algorithm for P_G. This is one of a group of three related questions that may be formalised as follows. (As before Σ is a finite alphabet of cardinality k.)

(a) Is there a *fully polynomial randomised approximation scheme* [17] (fpras) for estimating $P_G(k, \ldots, k)$? That is to say, is there a randomised algorithm that takes as input a group G and $\varepsilon > 0$, and produces as output a number Y (a random variable) such that

$$\Pr\left((1-\varepsilon)P_G(k, \ldots, k) \le Y \le (1+\varepsilon)P_G(k, \ldots, k)\right) \ge \frac{3}{4},$$

and, moreover, does so within time poly(n, ε^{-1})?

(b) Is there a polynomial-time *almost uniform sampler*[4] [16] for the orbits of Σ^n under the action of G? That is to say, is there a randomised algorithm that takes as input a group G and $\varepsilon > 0$, and produces as output a word $Y \in \Sigma^n$ (a random variable) such that for each orbit O,

$$(1-\varepsilon)N^{-1} \le \Pr(Y \in O) \le (1+\varepsilon)N^{-1},$$

where $N = P_G(k, \ldots, k)$ is the total number of orbits? The execution time is required to be bounded by poly(n, ε^{-1}).

(c) Is there a polynomial-time "almost-w" sampler for G, where the weight function $w : G \to \mathbb{N}$ is defined by $w(g) = k^{c(g)}$, where $c(g)$ denotes the number of cycles in g. That is to say, is there a randomised algorithm that takes as input a group G and $\varepsilon > 0$, and produces as output a permutation $Y \in G$ (a random variable) such that for each $g \in G$,

$$(1-\varepsilon)w(g)Z^{-1} \le \Pr(Y = g) \le (1+\varepsilon)w(g)Z^{-1},$$

where $Z = |G| \, P_G(k, \ldots, k)$? Again, the execution time is required to be bounded by poly(n, ε^{-1}).

The complexity of approximate counting and of almost uniform sampling are known to be closely related,[5] which would lead one to suppose that questions

---

[4]When this concept was first introduced, "generator" was used in place of "sampler", but the latter word is more specific.

[5]A rather precise statement of this relationship has been formulated by Jerrum, Valiant, and Vazirani [16].

(a), (b), and (c) ought to be equivalent. However, the situation here is atyp-
ical, and it is not clear, for example, whether resolving question (a) in the
affirmative would immediately settle either of the others. The two known
entailments are described in the following proposition, whose (routine) proof
may be found in [13, 14].

Proposition 3 *An affirmative answer to question* (c) *would entail affirmative answers to* (a) *and* (b).

I feel that questions (a)–(c) are quite significant, and they are all unresolved.
It is worth considering briefly why the proof technique (i.e., reduction) of
Theorem 1 cannot be used to provide a negative answer to question (a); after
all, there is convincing evidence (see [21, Thm 1.17]) that there is no fpras for
#INDSET, which is the starting point for the reduction. The catch is that the
reduction makes essential use of polynomial interpolation, and interpolation
does not preserve closeness of approximation.

Question (a) concerns the evaluation of P_G at points of the form $z_1 = z_2 =$
$\cdots = z_n = k$, for some positive integer k. While these are undoubtedly the
points of greatest combinatorial interest, it is natural to enquire whether it
is possible to approximate P_G at other points. Interestingly, relaxing either
constraint — that the z_i are assigned equal values, or that they are assigned
integer values — appears to lead to computational intractability. We explore
this phenomenon in the following subsection.

3.1   Negative results

The first result provides evidence that evaluating the cycle-index polynomial
at points whose coordinates are not all equal is computationally intractable.
Before presenting the result, we need to consider the nature of the evidence.
RP is the class of decision problems that can be solved in polynomial time
by a certain type of randomised algorithm which is allowed one-sided errors.
(See [3, p. 138] for a precise definition.) Rabin's celebrated primality test
is an example of such an algorithm. The class RP is widely regarded as
a reasonable extension to the class P, which preserves the notion of efficient
solvability. It would be surprising if all NP-complete problems were efficiently
solvable in the RP sense, and so it is conjectured that RP is strictly contained
in NP. Goldberg [9, 10] has shown the following.

Theorem 4 *Let* (a_i) *be a sequence of non-negative rationals, such that there
exists* i *with* $a_i > a_1^i$. *There can be no fpras for* CYCLEINDEX(a_i), *unless*
RP = NP.

The above theorem is couched in slightly different language to [9, Thm 23]
and [10, Thm 3], but the essential content is the same. The proof given in

those references is adequate to establish this version of the theorem. As an immediate corollary of Theorem 4 we have:

Corollary 5 *Assuming* RP $\neq$ NP, *there is no fpras for* CYCLEINDEX$(0, 2, 0, 2, \ldots)$.

In the light of the earlier discussion of the relationship between RP and NP, it seems unlikely that the cycle-index polynomial can be efficiently approximated at the point $0, 2, 0, 2, \ldots$. The combinatorial significance of Corollary 5 is that $P_G(0, 2, 0, 2, \ldots)$ counts the number of self-complementary orbits of the action of G on $\{0, 1\}^n$; an orbit is *self-complementary* if it is invariant under the interchange of letters 0 and 1. In this manner, the number of self-complementary graphs (graphs that are isomorphic to their own complements) may be expressed as an evaluation of the cycle-index polynomial.

Given the lack of progress on question (a), the following result of Goldberg and Jerrum may come as a surprise.

Theorem 6 *Let a be a positive rational that is not an integer. There can be no fpras for* CYCLEINDEX$(a, a, \ldots)$, *unless* RP = NP.

It is necessary to explain the phenomenon that allows the all important integer case to escape from the grasp of the proof of Theorem 6. Suppose a is as in the statement of Theorem 6, and let $h = \lceil a \rceil + 1$. Then it can be shown that

$$P_{\text{Alt}[h]}(a, \ldots, a) < P_{\text{Sym}[h]}(a, \ldots, a), \tag{3}$$

where Sym$[h]$ and Alt$[h]$ denote, respectively, the symmetric and alternating groups on $[h]$. In contrast, if a is a positive integer, inequality (3) must fail, as can be seen by considering the combinatorial interpretations of $P_{\text{Alt}[h]}(a, \ldots, a)$ and $P_{\text{Sym}[h]}(a, \ldots, a)$ in the integral case.

To exploit inequality (3) in the case that a is not an integer, we adapt the reduction of Theorem 1 as follows. Given a graph $\Gamma = (V, E)$ with n vertices and m edges, we again construct a permutation group G, this time of degree hm. The base set of G is $\Omega = E \times [h]$, and we start the construction of the generating set of G by adding for each $e \in E$, a set of permutations that generate the alternating group Alt$(\{e\} \times [h])$ on $\{e\} \times [h]$. As in the proof of Theorem 1 we introduce, for each vertex $v \in V$, a permutation g_v that acts on all sets $\{e\} \times [h]$ for which the edge e is incident at vertex v. The action of g_v on $\{e\} \times [h]$ is simply to transpose an arbitrary pair of elements.

As before there is a natural correspondence between subsets $U \subseteq V$ and objects in the group G; in this case, each vertex subset U corresponds not to a single element in G, but to a coset of $\prod_{e \in E} \text{Alt}(\{e\} \times [h])$ in G; conversely, each coset corresponds to two vertex subsets, U and $\overline{U}$, that are complements of each other. In other words, there is a bijection between cosets and bi-partitions $\{U, \overline{U}\}$ of the vertex set of Γ. Each bipartition $\{U, \overline{U}\}$ defines a *cut* containing all edges of Γ with one endpoint in U and one in $\overline{U}$. When the cycle-index polynomial of G is partitioned according to the cosets, by inequality (3), it is those cosets corresponding to maximum (cardinality) cuts that carry the greatest weight.

Now by employing the wreath product construction from the proof of Theorem 1, we may efficiently boost the weight of the cosets corresponding to maximum weight cuts until they account for almost all of the weight of the cycle-index polynomial. If this is done, then an approximation to the value of the cycle-index polynomial will yield the size of a maximum cut, and also an approximation to the number of such cuts. However, determining whether a graph Γ has a cut of cardinality at least C is an NP-complete problem [8, p. 210]. The upshot is that a fpras for CYCLEINDEX$(a, a, \ldots)$ would entail the existence of a fast randomised (RP-style) algorithm for a particular NP-complete problem, and hence for all problems in NP.

When a is an integer, one can have, at best, equality in (3). The reverse inequality is possible and corresponds to the "minimum cut" problem which, in the absence of further constraints, is trivial. It is hoped that the above account, though telegraphic, does give an intuitive feel for the phenomenon of intractability of the cycle-index polynomial at non-integer points.

3.2 Positive results

As we have observed, there is no known general and efficient procedure for estimating the cycle-index polynomial at integer points, or for sampling uniformly at random (u.a.r.) from the orbits of the action of G on Σ^n. However there is a promising approach to both of these problems, which will be described in this final subsection. We concentrate on the latter problem; any success that the method may find there easily translates to the former problem

Our approach to sampling orbits is to simulate an appropriately defined Markov chain. This technique has proved fruitful on a number of occasions in recent years; previous applications include an algorithm of Jerrum and Sinclair for estimating the permanent of a 0,1-matrix [15] and one of Dyer, Frieze, and Kannan for estimating the volume of a convex body in n-dimensional space [6]. In this instance we wish to construct a Markov chain $M = M(G, \Sigma)$ whose state space is Σ^n, and whose stationary distribution assigns equal probability to each orbit. In fact, we shall aim at something stronger, namely, a stationary distribution that assigns to each word $\alpha \in \Sigma^n$ a probability inversely proportional to the size of the orbit α^G containing α.

The transition probabilities from a state $\alpha \in \Sigma^n$ are specified by the following conceptually simple two-step experiment:

(1) choose g uniformly at random (u.a.r.) from the point stabiliser $G_\alpha = \{g \in G : \alpha^g = \alpha\}$;

(2) choose β u.a.r. from the set $\{\beta \in \Sigma^n : \beta^g = \beta\}$.

The new state is β.[6] Before analysing the stationary distribution of M, we should pause to consider the computational complexity of implementing the above experiment. Step (2) is computationally undemanding, and amounts to assigning, u.a.r. and independently, a symbol from Σ to each cycle of g. Step (1) is more interesting, and is equivalent (under randomised polynomial-time reducibility) to computing a setwise stabiliser in a permutation group, a task that includes deciding isomorphism of two graphs as a special case. The computational complexity of the setwise stabiliser problem is open: it is one of the very rare natural candidates for a problem that is in the class NP, but is neither in P nor NP-complete.

Although no general polynomial-time algorithm for implementing step (1) is known, and it is perfectly possible that none exists, there are significant classes of groups G for which step (1) does have an efficient implementation. Luks has shown that p-groups — groups in which every element has order a power of p for some prime p — is an example of such a class [18].

Returning to the Markov chain itself, we note immediately that M is ergodic, since every state can be reached from every other in a single transition, by selecting the identity permutation in step (1). The easiest way to get at the stationary distribution is perhaps by considering a random walk on the bipartite graph B that has vertex bipartition (Σ^n, G) and edge set $\{(\alpha, g) : \alpha^g = \alpha\}$. It is clear that the Markov chain M may be viewed as sampling the random walk on B after every even step. Let $\pi : \Sigma^n \to [0, 1]$ denote the stationary distribution of M. Then $\pi(x)$ is proportional to the degree of vertex α in the graph B, which is $|G_\alpha|$. It is an elementary group-theoretic fact that $|G_\alpha| \times |\alpha^G| = |G|$, and hence $\pi(\alpha)$ is inversely proportional to $|\alpha^G|$. We have therefore established:

Theorem 7 *Let π be the stationary distribution of Markov chain M. Then $\pi(\alpha) = |\alpha^G|^{-1} P_G(k, \ldots, k)^{-1}$ for all $\alpha \in \Sigma^n$; in particular, π assigns equal probability to each orbit α^G.*

Dually we might consider the Markov chain M' with state space G and transition probabilities modelled by an experiment in which steps (1) and (2) appear transposed. By relating M' to the random walk on B we easily obtain:

---

[6] Peter Cameron has observed that this process is defined for any group action, not just the special case of G acting on Σ^n by the permutation of positions.

Theorem 8 *Let π' be the stationary distribution of Markov chain M'. Then $\pi'(g) = k^{c(g)}|G|^{-1}P_G(k,\ldots,k)^{-1}$ for all $g \in G$, where $c(g)$ denotes the number of cycles in the permutation g.*

Note that the stationary distributions of the Markov chains M and M' match the sampling distributions specified in questions (b) and (c).

Although the topic of random walks on groups has received much attention (see for example the work of Aldous [1] and Diaconis [5]), previous authors have been concerned with walks which converge to a uniform distribution. The novel aspect of the current investigation is that the stationary distribution is required to be highly non-uniform.

We have seen that the stationary distribution of the Markov chain M is appropriate for sampling u.a.r. from the set of orbits of the G-action on Σ^n. For the Markov chain simulation approach to be computationally efficient, it is necessary for M to be *rapidly mixing*. We give a precise meaning to this informal requirement by insisting that M should be "close" to stationarity after a number of steps that is bounded by a polynomial in n. Since the size of the state space is exponential in n, this is a non-trivial requirement.

There are a number of ways of quantifying "closeness" to stationarity, but they are all essentially equivalent in this application. Consider an ergodic Markov chain with state space Q. Let $q \in Q$ be an arbitrary state, and denote by $P^t(q,\cdot)$ the distribution of the state at time t given that q is the initial state. Let π be the stationary distribution of M. Then the *variation distance* at time t with respect to the initial state q is defined to be

$$\delta_q(t) = \max_{S \subseteq Q} \left| P^t(q, S) - \pi(S) \right| = \tfrac{1}{2} \sum_{q' \in Q} \left| P^t(q, q') - \pi(q') \right|.$$

The rate of convergence may be measured by the function

$$\tau_q(\varepsilon) = \min\{t : \delta_q(t') \le \varepsilon \text{ for all } t' \ge t\}.$$

The Markov chain simulation technique will yield an efficient almost uniform sampler for orbits of the G-action on Σ^n provided the Markov chain $M(G,\Sigma)$ mixes rapidly. In recent years, techniques have been developed for bounding the mixing rate of combinatorially defined Markov chains, using, among other ideas, the relation between mixing rate and the expansion properties of the Markov chain viewed as a graph. Sinclair has provided a useful survey of these techniques, in addition to presenting some sharpened bounds [20]. Nevertheless, proofs of rapid mixing still tend to be technically involved.

Since no counterexamples have been identified, it remains a possibility that for any fixed alphabet Σ, the Markov chain $M(G,\Sigma)$ is rapidly mixing for all choices of G; specifically, that there is a *fixed* polynomial (in n and $\log \varepsilon^{-1}$)

that uniformly bounds $\tau_q(\varepsilon)$ for all possible groups G. If this were so — and the author would not hazard a conjecture on this point — then questions (b) and (c) would be answered in the affirmative, at least for groups for which the Markov chain could be efficiently simulated. Note that by Proposition 3, we would also have, indirectly, an affirmative answer to question (a).

A proof that the Markov chain $M(G, \Sigma)$ is rapidly mixing, even for the special case of G an abelian p-group, would have a significant algorithmic consequence in the field of statistical physics: it would imply the existence of a polynomial-time algorithm for estimating the partition function of a ferromagnetic p-state Potts system. We should therefore not be surprised if progress on the rapid mixing question for general permutation groups is rather slow. A more realistic programme for the short term is to produce a catalogue of families of groups G for which rapid mixing of $M(G, \Sigma)$ can be rigorously demonstrated. The following two results represent a modest start on this programme. Their proofs, which employ standard techniques, can be found in [14]. (The author is grateful to David Aldous for suggesting an improvement to the proof of Proposition 9, leading to a tighter bound on convergence time.)

Proposition 9 *Let* $\mathrm{Sym}[n]$ *denote the symmetric group on* $[n]$, *and* $M = M(\mathrm{Sym}[n], \Sigma)$ *the derived Markov chain defined earlier in the subsection. The convergence time of* M *is*

$$\tau_\alpha(\varepsilon) = \mathrm{O}(\log n\varepsilon^{-1}),$$

uniformly over the choice of initial state $\alpha \in \Sigma^n$.

Note that convergence is *very* rapid in the case of the full symmetric group. The constant implicit in the O-notation is dependent of the size of the alphabet Σ; for $\Sigma = \{0, 1\}$, the constant is 1, provided the logarithm is taken to base 2.

Proposition 10 *Let* G *denote any cyclic permutation group on* $[n]$, *and* $M' = M'(G, \Sigma)$ *the derived Markov chain defined earlier in the subsection. The convergence time of* M' *is*

$$\tau_g(\varepsilon) = \mathrm{O}(n \log n \log \varepsilon^{-1}),$$

uniformly over the choice of initial state $g \in G$.

The constant implicit in the O-notation is here independent of Σ.

Acknowledgement

I thank Leslie Goldberg for helpful comments on a draft of this article.

References

[1] David ALDOUS, Random walks on finite groups and rapidly mixing Markov chains, *Séminaire de Probabilités XVII 1981/82* (A. Dold and B. Eckmann, eds), Springer Lecture Notes in Mathematics **986**, pp. 243–297.

[2] László BABAI, Eugene LUKS, and Ákos SERESS, Fast management of permutation groups, *Proceedings of the 29th IEEE Symposium on Foundations of Computer Science*, Computer Society Press, 1988, pp 272–282.

[3] J. L. BALCÁZAR, J. DÍAZ, and J. GABARRÓ, *Structural Complexity, Volume I*, Springer-Verlag, Berlin, 1988.

[4] N. G. DE BRUIJN, Pólya's theory of counting, *Applied Combinatorial Mathematics*, E. F. Beckenbach (ed.), John Wiley and Sons, 1964, pp. 144–184.

[5] Persi DIACONIS, *Group representations in probability and statistics*, Institute of Mathematical Statistics, Hayward CA, 1988.

[6] Martin DYER, Alan FRIEZE, and Ravi KANNAN, A random polynomial time algorithm for approximating the volume of convex bodies, *Journal of the ACM* **38** (1991) pp. 1–17.

[7] Merrick FURST, John HOPCROFT, and Eugene LUKS, Polynomial time algorithms for permutation groups, *Proceedings of the 21st IEEE Symposium on Foundations of Computer Science*, Computer Society Press, 1980, pp. 36–41.

[8] Michael R. GAREY and David S. JOHNSON, *Computers and Intractability: A Guide to the Theory of NP-Completeness*, Freeman, San Francisco, CA, 1979.

[9] Leslie Ann GOLDBERG, *Efficient Algorithms for Listing Combinatorial Structures*, Cambridge University Press, 1993.

[10] Leslie Ann GOLDBERG, Automating Pólya theory: the computational complexity of the cycle index polynomial, *Information and Computation* **105** (1993), pp. 268–288.

[11] Frank HARARY and Edgar M. PALMER, *Graphical enumeration*, Academic Press, 1973.

[12] Mark JERRUM, A compact representation for permutation groups, *Journal of Algorithms* **7** (1986) pp. 60–78.

[13] Mark JERRUM, Uniform sampling modulo a group of symmetries using Markov chain simulation, *Expanding Graphs*, Joel Friedman (ed.), DIMACS Series in Discrete Mathematics and Theoretical Computer Science **10**, American Mathematical Society, 1993, pp. 37–47.

[14] Mark JERRUM, *Uniform sampling modulo a group of symmetries using Markov chain simulation.* Report ECS-LFCS-93-272, Department of Computer Science, University of Edinburgh, July 1993.

[15] Mark JERRUM and Alistair SINCLAIR, Approximating the permanent, *SIAM Journal on Computing* **18** (1989) pp. 1149–1178.

[16] M. R. JERRUM, L. G. VALIANT and V. V. VAZIRANI, Random generation of combinatorial structures from a uniform distribution, *Theoretical Computer Science* **43** (1986), pp. 169–188.

[17] Richard M. KARP and Michael LUBY, Monte-Carlo algorithms for enumeration and reliability problems, *Proceedings of the 24th IEEE Symposium on Foundations of Computer Science*, Computer Society Press, 1983, pp. 56–64.

[18] Eugene M. LUKS, Isomorphism of graphs of bounded valence can be tested in polynomial time, *Journal of Computer and System Sciences* **25** (1982), pp. 42–65.

[19] Charles C. SIMS, Computational methods in the study of permutation groups, *Computational Problems in Abstract Algebra*, J. Leech (ed.), Pergamon Press, New York, 1970, pp. 169–183.

[20] Alistair SINCLAIR, Improved bounds for mixing rates of Markov chains and multicommodity flow, *Combinatorics, Probability and Computing* **1** (1992), pp. 351–370.

[21] A. J. SINCLAIR, *Randomised algorithms, for counting and generating combinatorial structures*, Advances in theoretical computer science, Birkhäuser, Boston, 1993.

[22] L. G. VALIANT, The complexity of computing the permanent, *Theoretical Computer Science* **8** (1979), pp. 189–201.

[23] Leslie G. VALIANT, The complexity of enumeration and reliability problems, *SIAM Journal on Computing* **8** (1979), pp. 410–421.

Mixing of Random Walks
and Other Diffusions on a Graph

László Lovász[1] and Peter Winkler[2]

Abstract

We survey results on two diffusion processes on graphs: random walks and chip-firing (closely related to the "abelian sandpile" or "avalanche" model of self-organized criticality in statistical mechanics). Many tools in the study of these processes are common, and results on one can be used to obtain results on the other.

We survey some classical tools in the study of mixing properties of random walks; then we introduce the notion of "access time" between two distributions on the nodes, and show that it has nice properties. Surveying and extending work of Aldous, we discuss several notions of mixing time of a random walk.

Then we describe chip-firing games, and show how these new results on random walks can be used to improve earlier results. We also give a brief illustration how general results on chip-firing games can be applied in the study of avalanches.

1 Introduction

A number of graph-theoretic models, involving various kinds of diffusion processes, lead to basically one and the same issue of "global connectivity" of the graph. These models include: random walks on graphs, especially their use in sampling algorithms; the "avalanche" or "sandpile" model of catastrophic events, which is mathematically equivalent to "chip-firing" games; load balancing in distributed networks; and, somewhat more distantly but clearly related, multicommodity flows and routing in VLSI. In this paper we survey some recent results on the first two topics, as well as their connections.

Random walks. The study of random walks on finite graphs, a.k.a. finite Markov chains, is one of the classical fields of probability theory. Recently interest has shifted from asymptotic results to inequalities and other quantitative properties involving a finite, possibly even very small number of steps.

---

[1]Dept. of Computer Science, Yale University, New Haven, CT 06510
[2]AT&T Bell Laboratories 2D-147, Murray Hill, NJ 07974

Much of this was motivated by applications to computer science. Perhaps the most important of these (though certainly not the only one) is *sampling* by random walk (see, e.g., [21], [33], [25]). This method is based on the fact that (at least for connected non-bipartite undirected graphs, which is easy to guarantee), the distribution of the current node after t steps tends to a well-defined distribution π, called the *stationary distribution* (which is uniform if the graph is regular). So to draw an (approximately) uniformly distributed random element from a set V, it suffices to construct a regular, connected, non-bipartite graph on V, and run a random walk on this graph for a large fixed number of steps.

A good example to keep in mind is shuffling a deck of cards. Construct a graph whose nodes are all permutations of the deck, and whose edges lead from each permutation to those obtainable from a single shuffle. Then repeated shuffle moves correspond to a random walk on this (directed) graph.

A crucial issue for this algorithm is the choice of the number of steps. Informally, let as call the necessary number of steps the *mixing time*. The surprising fact, allowing these algorithmic applications, is that this mixing time may be much less than the number of nodes. For example, it takes only 7 moves [11] to shuffle a deck of 52 cards quite well, using the standard "dovetail" shuffle—even though the graph has 52! nodes. On an expander graph with n nodes, it takes only $O(\log n)$ steps to mix.

At the same time, proving good bounds on the mixing time even in quite special cases is a difficult question. Various methods have been developed for this. Using eigenvalues, it is easy to find the *mixing rate*, i.e., the quantity

$$\lim_{t \to \infty} d(\sigma^t, \pi)^{1/t},$$

where σ^t is the distribution of the node we are at after t steps, and d is the total variation (ℓ_1-)distance (or any other reasonable distance function). But this result does not tell the whole story for two reasons. First, the underlying graph in the cases of interest is exponentially large, (cf. the example of card shuffling), and the computation of the eigenvalues by the tools of linear algebra is hopeless. Second, the mixing rate tells us only the asymptotic behavior of the distance $d(\sigma^t, \pi)$ as $t \to \infty$, while we are interested in relatively small values of t (7 shuffle moves, for example). To be sure, eigenvalue methods can provide very sharp estimates, but for this, detailed information on the spectrum, and even on the eigenvectors, is needed (see Diaconis [21] or Chung and Yau [17]). This kind of spectral information can be derived, it seems, only in the presence of some algebraic structure, e.g. a large automorphism group.

Therefore, combinatorial techniques that yield only bounds on the mixing rate and mixing time are often preferable. Two main techniques that have been used are *coupling* and *conductance*. We only give a brief discussion of the second; see [35] for more details.

Recent work by the authors provides a further method to prove bounds on the mixing time. Our work was motivated by the following observation. There is no particular reason why a walk used in a sampling algorithm must be run for a fixed number of steps; in fact, more general stopping rules which "look where they are going" are capable of achieving the stationary distribution exactly, and just as fast. Motivated by this, we have studied stopping rules that achieve any given distribution, when starting from some other given distribution. It turns out that there is a surprising variety of such rules, many of which are optimal in the sense that they entail the smallest possible expected number of steps; one of them also minimizes the *maximum* number of steps. These rules are related to important parameters of the random walk, like hitting times and conductance. The expected number of steps in an optimal rule serves as a natural (non-symmetric) distance between the initial and final distributions.

The most important special case arises when one wishes to generate a node from the stationary distribution, starting from a given node. The "distance" from a node to the stationary distribution, maximized over all nodes, provides a precise natural definition of *mixing time* (considered by Aldous [5], [6] in the reversible case). This notion agrees, up to a constant factor, with most of the usual definitions of mixing time, which depend on a specific choice of how nearness to the limit distribution is measured.

In these considerations, we assume that the graph is known, and we put no restriction on the computation needed to decide when to stop. This requirement makes direct use of our stopping rules as sampling mechanisms unlikely. (We show in [38] that it is possible to obtain the exact stationary distribution with an unknown graph, not in as efficient a manner, although still in time polynomial in the maximum hitting time.) However, one can describe a simple rule whose implementation requires no knowledge about the graph other than its mixing time, takes only a constant factor more time, and yields a node whose distribution is approximately stationary. The machinery we build to determine the mixing time may thus be considered as a tool for analyzing this simple and practical sampling mechanism.

A main tool in the analysis of random walks on graphs is the Laplacian of the graph. Mixing times, hitting times, cover times and many other important parameters are closely related to the "eigenvalue gap" of this matrix, at least in the undirected case. A simpler but powerful tool is the "conservation equation" first noted by Pitman [42] (see Section 4).

Chip-firing and avalanches. Another diffusion process on graphs was introduced by Björner, Lovász and Shor [13] under the name of "chip-firing game". We place a pile of chips on each node of a directed graph, and then change this arrangement of chips as follows: we select a node which has at least as many chips as its outdegree, and move one chip from this node to each of its descendents. We call this step *firing* a node. This step is repeated as often as we wish or until no node remains that can be fired.

Procedures equivalent to chip-firing games were introduced, independently, at least three times (not counting the obvious similarity to neural nets, which remains unexplored). Engel [26], [27] considered a procedure he called the "probabilistic abacus", as a method of determining the limit distribution of certain Markov chains by combinatorial means. Spencer [47] introduced the special case when the underlying graph is a path, as a tool in analyzing a certain "balancing" game. In [4] Spencer's process was analyzed in greater detail. The analysis of the procedure was extended to general (undirected) graphs in [13], and to directed graphs by Björner and Lovász [14].

Chip-firing turns out to be closely related to the "avalanche" or "sandpile" model of catastrophic events (also called self-organized criticality), introduced by Bak, Tang and Wiesenfeld [12] and Dhar [20]. The nodes of the digraph represent "sites" where snow is accumulating. There is a special node, the "outside universe". Once the amount of snow on a site surpasses a given threshold, the site "breaks", sending one unit of snow to each of its out-neighbors, which in turn can break again etc., starting an avalanche. After some easy reductions, avalanches can be considered as chip-firing games; even firing the special node can be viewed as a snowfall.

A key property of these games is that from a given position, all sequences of firings behave similarly: either they all can be extended infinitely, or they all terminate after the same number of moves, with the same final position (Church-Rosser property). This was observed in [4] and in [20].

Considering a chip-firing process on a given digraph, we can ask a number of natural questions: will this procedure be finite or infinite? If finite, how long can it last? If infinite, how soon can it cycle? How many chips are needed for an infinite procedure? How does one determine if a given position (distribution of chips) can be transformed into another one by firings?

In the case of undirected graphs, these questions are more-or-less fully answered in [13], [14] and the work of Tardos [48]. For example, a finite procedure terminates in $O(n^4)$ steps; the shortest period of a periodic game is n; the minimum number of chips that allow an infinite game is m, the number of edges. There are polynomial time algorithms to determine if a position starts a finite or infinite game, and also to determine if two positions can be reached from each other. The case of directed graphs is more difficult, and the complexity of some of these questions is still open.

There is a strong connection between chip firings, random walks on graphs, and the Laplace operator. In particular, the "conservation equation" plays an important role. This connection in the undirected case was observed in

[13]; the extension to the directed case is due to [14], where it was used to show that no terminating firing sequence is longer than a polynomial times the length of the period of a periodic firing sequence. (This extends the result of [48], the directed case.) The new results on mixing times of random walks give improvements of these results. A converse inequality, conjectured in [14], will also be proved here, using the conservation equation.

There are a number of other diffusion processes on graphs, which we do not survey here in detail. *Load balancing* in distributed networks seems to be very closely related. In this model, every node of a (typically undirected and regular) graph corresponds to a processor, and each processor i is given a certain amount w_i of workload. The processors want to pass load to each other along the edges, so that eventually their loads should be (approximately) equal. This is quite similar in spirit to random walks on a regular graph, where "probability" is passed along the edges and eventually equalized. Indeed, upper and lower bounds on the time needed to equalize the loads ([2], [32]) involve parameters familiar from the theory of random walks: expansion rate, conductance, eigenvalue gap. On the other hand, there is a substantial difference: in chip-firing and random walks, load is distributed among the neighbors of a node evenly; in the load-balancing models, usually only one neighbor gets load in one step. Still, we hope that some of the ideas used in the analysis of random walks (and perhaps chip firing) might be applicable to a larger class of distribution processes.

2   Random walks, hitting and mixing times

Consider a strongly connected digraph $G = (V, E)$ with n nodes and m edges (we allow multiple edges and loops). We denote by a_{ij} or $a(i,j)$ the number of edges from i to j, and by d_i^- and d_i^+ the indegree and outdegree of node i, respectively. If the graph is undirected, then $d_v = d_v^+ = d_v^-$ is the degree of the node.

A *random walk* on G starts at a node w_0; if after t steps we are at a node w_t, we move to any node u with probability $a(w_t, u)/d^+(w_t)$. Clearly, the sequence of random nodes $(w_t : t = 0, 1, \ldots)$ is a Markov chain. The node w_0 may be fixed, but may itself be drawn from some initial distribution σ. We denote by σ^t the distribution of w_t:

$$\sigma_i^t = \Pr(w_t = i).$$

We denote by $M = (p_{ij})_{i,j \in V}$ the matrix of transition probabilities of this Markov chain. So

$$p_{ij} = \frac{a_{ij}}{d_i^+}.$$

The rule of the walk can be expressed by the simple equation

$$\sigma^{t+1} = M^{\mathsf{T}}\sigma^t,$$

(the distribution of the t-th point is viewed as a vector in $\mathbb{R}^V$), and hence

$$\sigma^t = (M^{\mathsf{T}})^t\sigma.$$

It follows that the probability p_{ij}^t that, starting at i, we reach j in t steps is given by the ij-entry of the matrix M^t.

If G is undirected (which is viewed as a special case of directed graphs, with each edge corresponding to a pair of arcs oriented in opposite directions) then this Markov chain is *time-reversible*. Roughly speaking, this means that every random walk considered backwards is also a random walk (see below for a precise definition). If, in addition, G is regular, then the Markov chain is *symmetric*: the probability of moving to u, given that we are at node v, is the same as the probability of moving to node v, given that we are at node u.

The probability distributions $\sigma^0, \sigma^1, \sigma^2, \ldots$ are of course different in general. We say that the distribution σ is *stationary* (or *steady-state*) for the graph G if $\sigma^1 = \sigma$. In this case, of course, $\sigma^t = \sigma$ for all $t \geq 0$. It is easy to see that there is a unique stationary distribution for every strongly connected digraph; we denote it by π. Algebraically, π is a left eigenvector of the transition matrix M, belonging to the eigenvalue 1.

A one-line calculation shows that for an undirected graph G, the distribution

$$\pi_i = \frac{d_i}{m} \tag{1}$$

is stationary (note that m is twice the number of undirected edges.) In particular, the uniform distribution on V is stationary if the graph is regular. An important consequence of this formula is that the stationary distribution is only a polynomial factor off the uniform (in terms of the number of edges, which we shall consider the input size of the graph. Loops and multiple edges are allowed.)

The stationary distribution for general directed graphs is not so easy to describe, but the following (folklore) combinatorial formula can be derived, e.g., from Tutte's "matrix-tree theorem". Let A_i denote the number of all spanning in-arborescences in G rooted at i. Then

$$\pi_i = \frac{d_i^+ A_i}{\sum_i d_i^+ A_i} \tag{2}$$

The stationary distribution on a directed graph can be very far from the uniform; it is easy to find examples where the stationary probability of some nodes is exponentially small (in the number of edges). The value

$$\hat{\pi} = \min_i \pi_i$$

is an important measure of how "lopsided" the walk is. However, if the digraph is eulerian, then the stationary distribution is proportional to the degrees just like in the undirected case:

$$\pi_i = \frac{d_i^+}{m}.$$

Specifically, the uniform distribution is stationary for every regular eulerian digraph.

The most important property of the stationary distribution is that if the digraph is aperiodic, i.e., the cycle lengths in G have no common divisor larger than 1, then the distribution of w_t tends to the stationary distribution, as $t \to \infty$. (This is not true if the cycle lengths have a common divisor, in particular, for undirected bipartite graphs.)

In terms of the stationary distribution, it is easy to formulate the property of time-reversibility of the random walk on an undirected graph: for every pair $i, j \in V$, $\pi_i p_{ij} = \pi_j p_{ji}$. This means that in a stationary walk, we step as often from i to j as from j to i. From (1), we have $\pi_i p_{ij} = 1/m$ if $ij \in E$, so we see that we move along every edge, in every given direction, with the same frequency. If we are sitting on an edge and the random walk just passed through it, then the expected number of steps before it passes through it in the same direction again is m.

There is a similar fact for nodes, valid for all digraphs: if we are sitting at a node and the random walk just visited this node i, then the expected number of steps before it returns is $1/\pi_i$. If G is a regular eulerian digraph (in particular, a regular undirected graph), then this "return time" is just n, the number of nodes.

The *mixing rate* is a measure of how fast the random walk converges to its limiting distribution. This can be defined as follows. If the digraph is aperiodic, then $p_{ij}^{(t)} \to d_i/(2m)$ as $t \to \infty$, and the mixing rate is

$$\mu = \limsup_{t \to \infty} \max_{i,j} \left| p_{ij}^{(t)} - \frac{d_i}{2m} \right|^{1/t}.$$

One could define the notion of "mixing time" as the number of steps before the distribution of w_t will be close to uniform (how long should we shuffle a deck of cards?). This number will be about $(\log n)/(1-\mu)$. However, the exact value depends on how (in which distance) the phrase "close" is interpreted. Another concern is that this definition excludes periodic digraphs, and is very pessimistic in the case of "almost periodic" digraphs. For example, if G is obtained from a complete bipartite graph by adding an edge, then after a single step the distribution will alternate between almost uniform on one color class, and the other, but it takes a (relatively) long time before this alternation disappears. In applications to sampling, simple averaging tricks take care of this problem. Soon we will be able to introduce a more sophisticated, but "canonical" definition of mixing time.

In this paper, we do not study other important parameters of random walks, like cover times, commute times and the like. But one "time" will play an important role in the analysis of mixing speed: the *hitting time* (or *access time*) $H(i, j)$ is the expected number of steps before node j is visited, starting from node i. We denote by $H(G)$ the largest hitting time between any two nodes of the graph G. For undirected graphs, hitting times are polynomial in the number of edges ([1]). Brightwell and Winkler [15] proved that for every simple graph, $H(G) \leq (4/27)n^3$, and determined the graph that provides the maximum.

For digraphs, hitting times are not bounded by any polynomial of the number of edges in general. In fact, they are closely tied to the smallest stationary probability $\hat{\pi}$. Björner and Lovász proved in [14] that

$$H(G) \leq \sum_{i \in V} \frac{d_i^+}{\pi_i}, \tag{3}$$

which, together with the trivial lower bound, implies that

$$\frac{1}{\hat{\pi}} - 1 \leq H(G) \leq \frac{m}{\hat{\pi}}. \tag{4}$$

Hitting times have many interesting combinatorial and algebraic properties; see [35] for several of these. We only state here two special properties, for later reference. The *random target identity* states that

$$\sum_j \pi_j H(i, j) = C \tag{5}$$

is independent of the choice of i; in other words, the expected number of steps we have to walk to hit a node randomly chosen from the stationary distribution is C, independent of the starting point (see, e.g., the "right averaging principle" in Aldous [5]).

The hitting time from i to j may be different from the hitting time from j to i, even in an undirected regular graph. Still, one expects that time-reversibility should give some sort of symmetry of these quantities. One symmetry property of hitting times for undirected graphs was discovered by Coppersmith, Tetali and Winkler [19]:

$$H(i, j) + H(j, k) + H(k, i) = H(i, k) + H(k, j) + H(j, i) \tag{6}$$

for every three nodes.

3   Mixing, eigenvalues and conductance

In this section we give a brief account of the use of these two tools in estimating the speed of mixing of a random walk. A more detailed survey, at least in the case of undirected graphs, can be found in [35].

The matrix M has eigenvalue 1, with corresponding left eigenvector π and corresponding right eigenvector $\mathbf{1}$, the all-1 vector on V. It follows from the Frobenius-Perron Theorem that every other eigenvalue λ satisfies $|\lambda| \leq 1$ and if G is non-periodic, then in fact $|\lambda| < 1$. We denote by μ the largest absolute value of any eigenvalue different from 1.

Now the key fact in the use of eigenvalue techniques is the following. Let σ be any starting distribution, then

$$\sigma^t - \pi = (M^T)^t(\sigma - \pi)$$

and hence it is easy to derive the following:

Theorem 3.1 *For every starting distribution σ, every $t \geq 1$ and $A \subseteq V$,*

$$|\sigma^t(A) - \pi(A)| \leq \frac{1}{\sqrt{\hat{\pi}}}\mu^t.$$

Conductance. Let G be a digraph and $S \subset V$, $S \neq \emptyset$. Let $e(S,T)$ denote the number of edges connecting a set S to a set T. We define the *conductance* of the set $S \subset V$, $S \neq \emptyset$ by

$$\Phi(S) = \frac{1}{\pi(S)\pi(V \setminus S)} \sum_{i \in S} \pi(i) \frac{e(i, V \setminus S)}{d_i^+}$$

and the conductance of the graph by

$$\Phi = \min_S \Phi(S),$$

where the minimum is taken over all non-empty proper subsets $S \subset V$. If the graph is a d-regular and undirected, then the conductance of S is

$$\Phi(S) = \frac{n}{d} \frac{e(S, V \setminus S)}{|S| \cdot |V \setminus S|},$$

which is (up to normalization) the edge-density in the cut determined by S.

To digest this quantity a little, note that $\sum_{i \in S} \pi(i)e(i, V \setminus S)/d_i^+$ is the frequency with which a stationary random walk switches from S to $V \setminus S$; while $\pi(S)\pi(V \setminus S)$ is the frequency with which a sequence of independent random elements of V, drawn from the stationary distribution π, switches from S to $V \setminus S$. So Φ can be viewed as a certain measure of how independent consecutive nodes of the random walk are.

Sinclair and Jerrum [33] established a connection between the spectral gap and the conductance of an undirected graph. A similar result for the related, but somewhat different parameter called *expansion rate* was proved by Alon [3] and, independently, by Dodziuk and Kendall [23] (cf. also Diaconis and Stroock [24]). All these results may be considered as discrete versions of Cheeger's inequality in differential geometry.

Theorem 3.2 *If G is an undirected graph, then every eigenvalue $\lambda \neq 1$ of M satisfies*

$$\lambda \leq 1 - \frac{\Phi^2}{8}.$$

This result allows an eigenvalue near -1, which means that the graph is almost bipartite. While such an eigenvalue prevents us from applying 3.1 right away, it is in fact easy to handle. For example, we may attach d_i loops at each node i; for the random walk on this modified graph we get

Corollary 3.3 *For any starting distribution σ, any $A \subseteq V$ and any $t \geq 0$,*

$$\left| \sigma^t(A) - \pi(A) \right| \leq \frac{1}{\sqrt{\hat{\pi}}} \left(1 - \frac{\Phi^2}{8} \right)^t.$$

See Diaconis and Stroock [24], Mihail [41], Fill [29], Sinclair [45], and also Lovász and Simonovits [36] for sharper bounds, connections with multicommodity flows, and for extensions to the directed case.

4   Stopping rules and exit frequencies

Examples. There are several examples of "stopping rules" that can achieve specified distributions in an elegant or surprising manner. We consider two; several more are mentioned in [39].

Consider the following interesting fact from folklore. Let G be a cycle of length n and start a random walk on G from a node u. Then the probability that v is the last node visited (i.e., the a random walk visits every other node before hitting v) is the same for each $v \neq u$.

While this is not an efficient way to generate a uniform random points of the cycle, it indicates that there are entirely different ways to use random walks for sampling than walking a given number of steps. This particular method does not generalize; in fact, apart from the complete graph, the cycle is the only graph which enjoys this property (see [37]).

Consider another quite simple graph, the cube, which we view as the graph of vertices and edges of $[0,1]^n$. Let us do a random walk on it as follows: at each vertex, we select an edge incident with the vertex at random, then flip a coin. If we get "heads" we walk along the edge; if "tails" we stay where we are. We stop when we have selected every direction at least once (whether or not we walked along the edge).

It is trivial that after we have selected an edge in a given direction, the corresponding coordinate will be 0 or 1 with equal probability, independently of the rest of the coordinates. So the vertex we stop at will be uniformly distributed over all vertices.

This method takes about $n \ln n$ coin flips on the average, thus about $\frac{1}{2} n \ln n$ actual steps, so it is a quite efficient way to generate a random vertex of the cube, at least if we insist on using random walks (of course, to choose the coordinates independently is simpler and faster). We will see that it is in fact optimal.

Stopping rules. To begin a systematic study of stopping rules, we first define them. A *stopping rule* Γ is a map that associates with every walk w in the digraph G a number $0 \le \Gamma(w) \le 1$. We interpret $\Gamma(w)$ as the probability of continuing given that w is the walk so far observed, each such stop-or-go decision being made independently. We can also regard Γ as a random variable with values in $\{0, 1, \ldots\}$, whose distribution depends only on the $w_0, \ldots, w_\Gamma$; thus we stop at w_Γ.

The *mean length* $\mathrm{E}\Gamma$ of the stopping rule Γ is its expected duration; if $\mathrm{E}\Gamma < \infty$ then with probability 1 the walk eventually stops, and thus σ^Γ is a probability distribution. A stopping rule Γ for which $\sigma^\Gamma = \tau$ is also called a *stopping rule from σ to τ*.

For any strongly connected digraph G and any distribution τ on $V(G)$, there is at least one finite stopping rule Γ such that $\sigma^\Gamma = \tau$; namely, we select a target node j in accordance with τ and walk until we reach j. We call this the "naive" stopping rule $\Omega_{\sigma\tau}$. Obviously, the mean length of $\Omega_{\sigma\tau}$ is given by

$$\mathrm{E}\Omega_{\sigma\tau} = \sum_{i,j} \sigma_i \tau_j H(i,j).$$

In the case when $\tau = \pi$ is the stationary distribution, this formula can be simplified using the "random target identity" (5), and we get that the mean length of the naive rule to reach π is C, independently of the starting distribution.

We often think of a stopping rule Γ as a means of moving from a starting distribution σ to a given target distribution $\tau = \sigma^\Gamma$. Such a Γ is said to be *mean-optimal* or simply *optimal* (for σ and τ) if $\mathrm{E}\Gamma$ is minimal. The mean length of a mean-optimal stopping rule from σ to τ will be denoted $H(\sigma, \tau)$. We call this number the *access time* from σ to τ, and think of it as a generalized hitting time.

Trivially, $H(\sigma, \tau) = 0$ if and only if $\sigma = \tau$. It is easy to see that the following *triangle inequality* is satisfied for any three distributions σ, ρ and τ:

$$H(\sigma, \tau) \le H(\sigma, \rho) + H(\rho, \tau). \tag{7}$$

(To generate τ from σ, we can first use an optimal rule to generate ρ from σ and then use the node obtained as a starting node for an optimal rule generating τ from ρ). We should warn the reader, however, that $H(\sigma, \tau) \ne H(\tau, \sigma)$ in general.

We have seen that the access time $H(\sigma, \tau)$ has the properties of a metric on the space of node-distributions, except for symmetry; the latter is of course too much to expect since the ordinary hitting time, even for an undirected graph, is not generally symmetric.

Clearly if τ is concentrated at j (for which we write, rather carelessly, "$\tau = j$") then

$$H(\sigma, j) = \sum_i \sigma_i H(i, j), \tag{8}$$

since the only optimal stopping rule in this case is Ω_j, "walk until node j is reached."

By considering the naive rule Ω_τ, we get the inequality

$$H(\sigma, \tau) \le \sum_{i,j} \sigma_s \tau_j H(i, j) . \tag{9}$$

This may be quite far from equality; for example, $H(\sigma, \sigma) = 0$ for any σ.

We set $H_{\max}(\tau) = \max_\sigma H(\sigma, \tau) = \max_i H(i, \tau)$. From the point of view of applications, stopping rules generating nodes from the stationary distribution are of particular interest. The value $T_{\mathrm{mix}} = \max_i H(i, \pi)$ (the mean time of an optimum rule, starting from the worst point) is a natural and very useful definition of the mixing time.

It turns out that for given target distribution τ there are at least four interesting optimal stopping rules: the *filling rule*, the *local rule*, the *shopping rule* and the *threshold rule*. We describe these rules, together with some important non-optimal stopping rules, a bit later.

The conservation law. Let us now fix the digraph, a starting distribution σ and a finite stopping rule Γ. The expected number x_i of times the walk leaves node i before stopping will be called the *exit frequency* of node i for Γ. Clearly

$$\mathrm{E}\Gamma = \sum_i x_i.$$

Exit frequencies were considered by Pitman [42]; he gave the following simple but very powerful "conservation law", relating them to the starting and ending distributions:

Lemma 4.1 *The exit frequencies of any stopping rule from σ to τ satisfy the equation*

$$\sum_i p_{i,j} x_i - x_j = \tau_j - \sigma_j .$$

The identity expresses the simple fact that the probability of stopping at node j is the expected number of times j is entered minus the expected number of times j is left. The first application of this identity is the following theorem ([39]), relating different rules leading from the same starting distribution to the same target distribution:

Theorem 4.2 *Fix σ and let Γ and Γ' be two finite stopping rules from σ to τ with exit frequencies x and x' respectively. Let $D = \mathbb{E}\Gamma - \mathbb{E}\Gamma'$ be the difference between their mean lengths. Then $x' - x = D\pi$.*

It follows from Theorem 4.2 that the exit frequencies of any mean-optimal stopping rule from σ to τ are the same. We denote them by $x_i(\sigma, \tau)$.

Let us determine the exit frequencies in some simple cases. The first result is from Aldous [5]. Several related formulas could be derived using relations to electrical networks, as in [18] or [48].

Lemma 4.3 *The exit frequencies $\tilde{x}$ for the naive stopping rule Ω_j in reaching node j from node i are given by*

$$\tilde{x}_k = \pi_k(H(i,j) + H(j,k) - H(i,k)).$$

More generally, the exit frequencies for the naive stopping rule Ω_τ from initial distribution σ are given by

$$\tilde{x}_k = \pi_k \sum_{i,j} \sigma_i \tau_j (H(i,j) + H(j,k) - H(i,k)) = \pi_k (\mathbb{E}\Omega_{\sigma\tau} + H(\tau,k) - H(\sigma,k)) .$$

Combining this lemma with Theorem 4.2, we get the following general formula for exit frequencies:

Theorem 4.4 *The exit frequencies of a mean-optimal stopping rule from σ to τ are given by*

$$x_k(\sigma, \tau) = \pi_k \left(H(\sigma, \tau) + H(\tau, k) - H(\sigma, k) \right).$$

Any node j for which $x_j = 0$ is called a *halting node*. By definition we stop immediately if and when any halting node is entered. (But of course we may stop in other nodes too, just not all the time.) The following theorem from [39] gives an extremely useful characterization of optimality.

Theorem 4.5 *A stopping rule* Γ *is optimal if and only if it has a halting node.*

The "if" part is a trivial consequence of Theorem 4.2. The "only if" part is more difficult: we have to prove that from every σ to every τ there is a stopping rule that has a halting node. There are several ways to specify such a rule. Later on we shall describe four optimal stopping rules. Any of these could be used to prove this theorem, but none of the proofs is really straightforward, and we don't give any of them here.

This theorem shows that from the two stopping rules on the cycle and the cube, discussed as introductory examples, the first is not optimal, but the second is (the node of the cube opposite the starting node is a halting node).

From Theorem 4.5, a formula for the access times follows easily. Consider an optimum stopping rule from σ to τ. Let j be any node, and consider the triangle inequality:

$$H(\sigma, j) \leq H(\sigma, \tau) + H(\tau, j)$$

The right hand side can be interpreted as the expected number of steps in a stopping rule that consists of first following an optimal rule from σ to τ and then following the naive rule (which is clearly the only optimal rule) from τ to j. Now if j is the halting node of the optimum rule from σ to τ then, trivially, it is a halting node for this composite rule, and so the composite rule is optimal. Thus for at least one j, equality holds. Rearranging, we get that

$$H(\sigma, \tau) = \max_j (H(\sigma, j) - H(\tau, j)). \tag{10}$$

Note that the access times on the right hand side can be expressed by the hitting times, using (8):

$$H(\sigma, \tau) = \max_j \sum_i (\sigma_i - \tau_i) H(i, j).$$

There is, in fact, a more general formula for the exit frequencies, which can be derived by similar arguments:

$$x_k(\sigma, \tau) = \pi_k \left(\sum_i (\tau_i - \sigma_i) H(k, i) - \min_j \sum_i (\tau_i - \sigma_i) H(j, i) \right)$$

In the special case of undirected graphs and target distribution π (which is perhaps the most common in applications of random walk techniques to sampling), we can use the cycle-reversing identity (6) and the random target identity (5) to obtain the following formula for the exit frequencies of an optimal rule:

$$x_k = \pi_k (\max_j H(j, i) - H(k, i)) \tag{11}$$

and

$$H(i, \pi) = \max_j H(j, i) - H(\pi, i) . \tag{12}$$

We have thus identified the halting node in an undirected graph, in attaining the stationary distribution from node i, as the node j *from which* the hitting time to i is greatest. This seems slightly perverse in that we are interested in getting from i to j, not the other way 'round!

Examples. Consider the classic case of a random walk on the path of length n, with nodes labeled $0, 1, \ldots, n$. We begin at 0, with the object of terminating at the stationary distribution.

The hitting times from endpoints are $H(0, j) = H(n, n - j) = j^2$ and the stationary distribution is

$$\pi = (\frac{1}{2n}, \frac{1}{n}, \frac{1}{n}, \ldots, \frac{1}{n}, \frac{1}{2n}).$$

The naive stopping rule has a halting node, namely n, and hence it is optimal. From (11) we have

$$x_k = \pi_k(H(n, 0) - H(k, 0)) = \pi_k H(n, k) = \pi_k(n - k)^2$$

and

$$H(0, \pi) = \sum_{i=0}^{n} x_k = \frac{1}{2n} n^2 + \sum_{i=1}^{n-1} \frac{1}{n}(n - k)^2 = \frac{n^2}{3} + \frac{1}{6} .$$

From this it is not difficult to derive that for a cycle of length n, n even,

$$H(s, \pi) = \frac{n^2}{12} + \frac{1}{6}$$

as compared with expected time

$$\frac{n - 1}{n} \frac{n(n - 1)}{2} = \frac{(n - 1)^2}{2}$$

for staying at 0 with probability $1/n$ else walking until the last new vertex is hit as in the example discussed earlier.

The random walk on the following digraph is sometimes called the *winning streak*. Let $V = \{0, 1, \ldots, n - 1\}$, and connect i to $i + 1$ by an edge for $i = 0, \ldots n - 2$; also connect i to 0 for $i = 1, \ldots, n - 1$. It is easy to check that the stationary distribution is

$$\pi_i = \frac{2^{n-i-1}}{2^n - 1}.$$

Hence the exit frequencies x_i for an optimal rule from 0 to π can be determined using Lemma 4.1, working backwards from $i = n - 1, n - 2, \ldots$, obtaining

$$x_i = (n - i - 1)\frac{2^{n-i-1}}{2^n - 1}.$$

Summing over all nodes, we get

$$H(0, \pi) = \frac{(n - 2)2^n + 2}{2^n - 1} = n - 2 + O(n2^{-n}).$$

Next we describe four optimal stopping rules.

The filling rule. This rule is the discrete version of the "filling scheme," introduced by Chacon and Ornstein [16] and shown by Baxter and Chacon [10] to minimize expected number of steps. We call it the *filling rule* (from σ to τ), and define it recursively as follows. Let p_i^k be the probability of being at node i after k steps (and thus not having stopped at a prior step); let q_i^k be the probability of stopping at node i in *fewer* than k steps. Then if we are at node i after step k, we stop with probability $\min(1, (\tau_i - q_i^k)/p_i^k)$.

Thus, the filling rule stops myopically as soon as it can without overshooting the target probability of its current node. One can prove that it is a finite stopping rule and thus it does in fact achieve τ when started at σ. One can also prove that it has a halting node.

The filling rule has a "now-or-never" property that once a node is exited, we never stop there later. In fact, it can be described in terms of "deadlines" g_i: we stop at node j if we hit it *before* this time; if we hit the node j at time t where $g_j < t \leq g_j + 1$, then we stop with probability $g_j + 1 - t$; we don't stop if we hit it after time $g_j + 1$. A halting node j gets $g_j = \infty$.

The threshold rule. Every "threshold vector" $h = (h_1, \ldots, h_n)$, $h_i \in [0, \infty]$ gives rise to a stopping rule in a manner opposite to the "deadlines" mentioned in connection with the filling rule: we stop at node j if we hit it *after* time $h_j + 1$; if we hit the node j at time t where $h_j < t \leq h_j + 1$, then we stop with probability $t - h_j$; we don't stop if we hit it before time h_j. A rule obtained this way is called a *threshold rule*.

The threshold vector may not be uniquely determined by a threshold rule (e.g. all possible thresholds h_i smaller than the time before any possible walk reaches i are equivalent), but by convention we always consider the vector each of whose coordinates is minimal. Then in view of Theorem 4.5, the threshold rule is optimal just when some coordinate of the associated threshold vector is zero.

Theorem 4.6 *For every target distribution τ there is a mean-optimal threshold rule.*

The threshold rule has a couple of further properties that distinguish it among other rules. First, if τ has sufficient support then it is bounded:

Theorem 4.7 *Suppose that every directed cycle contains a node i with $\tau_i > 0$. Then there is a $K > 0$ such that, for every starting distribution, the threshold rule always stops in fewer than K steps.*

The condition of Theorem 6 is also necessary in the sense that if a cycle with target probability 0 exists, then starting at a node of this cycle, no bound can be given on the number of steps in the threshold (nor on the number of steps of any other stopping rule).

The threshold rule is special among all rules in the following sense:

Theorem 4.8 *The maximum number of steps taken by the threshold rule is not larger than the maximum number of steps taken by any other rule from the same starting distribution to the same target.*

The local rule. Let x_i be the exit frequencies for an optimal stopping rule from σ to τ, i.e., solutions of the conservation law with $\min_i x_i = 0$. (An easy algebraic argument shows that for any σ and τ, there is a unique solution of the conservation equation with this property.) Consider the following "local" rule: if we are at node i, we stop with probability $\tau_i/(x_i + \tau_i)$, and move on with probability $x_i/(x_i + \tau_i)$ (if $x_i + \tau_i = 0$ the stopping probability does not need to be defined). Thus the probability of stopping depends only on the current node, not the time.

One can prove that the local rule generates τ. It is mean-optimality is clear since the node j with $x_j = 0$ is a halting node.

The shopping rule. Any probability distribution on the *subsets* of the node set V provides a stopping rule: "choose a subset U from ρ, and walk until some node in U is hit." The naive rule is of course a special case, with ρ concentrated on singletons. The special case when ρ is concentrated on a chain of subsets is more efficient:

Theorem 4.9 *For every target distribution τ, there exists a unique distribution ρ which is concentrated on a chain of subsets and gives a stopping rule for generating τ. This stopping rule is optimal.*

The chain supporting the distribution ρ can be constructed recursively, starting from V and going down. Once we know that such a rule from σ to τ exists, its optimality is obvious, since a node in the smallest member of the chain is never exited.

Another rather neat way to think of this rule is to assign the real "price" $r(i) = \sum\{\rho(U) : i \in U\}$ to each node i. The "shopping rule" is then implemented by choosing a random real "budget" r uniformly from $[0,1]$ and walking until a node j with $r(j) \leq r$ is reached.

The shopping rule shares with the filling rule Φ the "now-or-never" property that once a node is exited, it can never be the node at which the rule stops.

It is interesting to point out that the four stopping rules described above have a lot of common features. Of course, they all have the same exit frequencies and halting nodes. Each is described in terms of a numerical vector on V (deadlines, thresholds, exit frequencies, prices). Each of these vectors can be calculated from the starting and target distribution, by an algorithm that is polynomial in the number of nodes (which is unfortunately not good enough in a typical application of these techniques to sampling, where the number of nodes is exponential).

Each of these rules (or the corresponding vector) defines an ordering (with ties—technically, a "preorder") of the nodes for every σ and τ. These orderings are in general different.

On the other hand, the four rules described above are different, and have in fact quite different properties. The threshold rule is bounded if, say, the target distribution has full support; the other three are, in general, not. The filling and shopping rules have the "now or never" property, but the other two rules do not. Finally, the filling rule has the "inverse boundedness" property that there is a time K so that it never stops after time K except in a halting node, which is not shared by any of the others.

5 Mixing times

We can define *the mixing time* of a random walk as $T_{\mathrm{mix}} = \max_s H(s, \pi)$. This is not quite in line with the usual definition of mixing time, which is the smallest t such that, for every initial distribution σ, the distribution σ^t of the t-th element is "close" to π in one sense or another. To be specific, say we want $\sigma_i^t \geq (9/10)\pi_i$ for all i. (In [37], the dependence on a parameter c in place of $9/10$ is also studied, but here we simplify our discussion by fixing this value.)

It is not immediately clear how to compare these two definitions. On the one hand, the traditional definition requires only approximate mixing, so it could be much less than our mixing time. On the other hand, the traditional definition is restricted to a trivial stopping rule (stop after t steps), and so it could be lead to much larger stopping times.

To be precise, we have to make one more point. If the graph is periodic (i.e., the lengths of its cycles have a common divisor larger than 1, say we have a bipartite graph), then σ^t may never be close to π. The way out is to do some kind of averaging: the (somewhat improperly named) "continuous time" model corresponds to choosing t from a Poisson distribution, while the "lazy walk" trick (see e.g. Lovász and Simonovits [36] corresponds to choosing t from a binomial distribution.

It turns out that none of these differences mean too much, at least if we allow averaging. In fact, the following value is a lower bound on both versions of mixing time; on the other hand, both versions are at most a constant factor larger.

Let T_{fill} denote the smallest T such that for every starting distribution σ, there is a stopping rule Φ with mean length at most T such that $\sigma_i^\Phi \geq (9/10)\tau_i$ for all i. We can modify this definition, by using a different notion of approximation, the so-called total variation distance: let T_{tv} denote the smallest T such that for every starting distribution σ, there is a stopping rule Φ with mean length at most T such that $|\sigma^\Phi(A) - \pi(A)| \leq 1/10$ for every set A of nodes.

Obviously,

$$T_{\text{tv}} \leq T_{\text{fill}} \leq T_{\text{mix}}.$$

The last two quantities are always close to each other (this is a consequence of a simple folklore argument):

Theorem 5.1
$$T_{\text{mix}} \leq \frac{10}{9} T_{\text{fill}}.$$

On the other hand, T_{tv} and T_{fill} may be far apart, as the "winning streak" example shows. Aldous (see e.g [5]) proved a converse inequality in the time-reversible case. Adapted to our case (and improving the constant a little), this implies:

Theorem 5.2 *If the graph G is undirected, then*

$$T_{\text{mix}} \leq 4 T_{\text{tv}}.$$

In the general case, the following inequality can be proved (the "winning streak" graph shows that it is tight).

Theorem 5.3
$$T_{\text{mix}} \leq O(\log(1/\hat{\pi})) T_{\text{tv}}.$$

Now we turn to the issue of how to implement optimal or near-optimal rules, to generate a node from the stationary distribution. It turns out that there exist simple, easily implementable rules that give a good approximation of the stationary distribution, while having a mean length only a constant factor more than the mixing time.

The *uniform averaging rule* $\Phi = \Phi_t$ ($t \geq 0$) is defined as follows: choose a random integer Y uniformly from the interval $0 \leq Z \leq t - 1$, and stop after Y steps. (To describe this as a stopping rule: stop after the j-th step with probability $1/(t - j)$, $j = 0, \ldots, t-1$).

Theorem 5.4 *Let Y be chosen uniformly from $\{0, \ldots, t\}$. Then for any starting distribution σ, any $0 \leq c \leq 1$, and any $A \subseteq V$,*

$$|\sigma^Y(A) - \pi(A)| \leq \frac{1}{t}H(\sigma, \pi).$$

In particular, if $t \geq (1/\varepsilon)T_{\mathrm{mix}}$ then

$$|\sigma^Y(A) - \pi(A)| \leq \varepsilon.$$

The contents of this (rather simple) theorem is that the averaging rule does as well as any sophisticated stopping rule, at least up to an arbitrarily small error and a constant factor in the running time. To illustrate how an "arbitrary" stopping rule can be related to the averaging rule, we sketch the proof.

Let Ψ be an optimal stopping rule from σ to π. Consider the following rule: follow Ψ until it stops at v^Ψ, then generate $Z \in \{0, \ldots, t-1\}$ uniformly and independently from the previous walk, and walk Y more steps. Since Ψ stops with a node from the stationary distribution, $\sigma^{\Psi+t}$ is also stationary for every $t \geq 0$ and hence so is $\sigma^{\Psi+Z}$.

On the other hand, let $Y = \Psi + Z \pmod{t}$, then Y is uniformly distributed over $\{0, \ldots, t - 1\}$, and so

$$\begin{aligned}
\sigma_i^Y &= \Pr(v^Y = i) \geq \Pr(v^{\Psi+Z} = i) - \Pr(v^{Psi+Z} = i, \; v^Z \neq i) \\
&= \sigma^{\Psi+Z} - \Pr(v^{\Psi+Z} = i, \; \Psi + Z \geq t).
\end{aligned}$$

Hence for every set A of states,

$$\pi(A) - \sigma^Y(A) = \pi(A) - \sigma^{\Psi+Z}(A) + \Pr(v^{\Psi+Z} \in A, \; \Psi+Z \geq t) \leq \Pr(\Psi+Z \geq t).$$

Now for any fixed value of Ψ, the probability that $\Psi + Z \geq t$ is at most Ψ/t, and hence

$$\Pr(\Psi + Z \geq t) \leq \mathrm{E}(\Psi/t) = \frac{H(\sigma, \pi)}{t},$$

which proves the theorem.

Theorem 5.4 only asserts closeness in the total variation distance, not pointwise. Also, one would like that the error diminishes faster: it should be enough to choose t proportional to $\log(1/\varepsilon)$ rather than proportional to $1/\varepsilon$. We can give a slightly more complicated rule that satisfies these requirements.

Choose $M = \lceil \log \varepsilon \rceil$, and let Y be the sum of M independent random variables, each being uniform over $\{0, \ldots, \lceil 8T_{\text{mix}} \rceil\}$. Clearly $EY \approx 4T_{\text{mix}} \log(1/\varepsilon)$. Furthermore, stopping after Y steps gives a distribution very close to the stationary:

Theorem 5.5 *For any starting distribution σ,*

$$\sigma^Y \geq (1 - \varepsilon)\pi.$$

(One little drawback in comparison with Theorem 5.4: we have to use the worst-case bound on the mixing time, not the access time from the given starting distribution.)

Blind rules. The averaging rules discussed above have an important property, which makes them practical but, at the same time, somewhat contrary to the philosophy of intelligent stopping rules: they don't look where they are. More exactly, let us call a stopping rule Γ *blind* if $\Gamma(w)$ depends only on the length of the walk w. Another way of describing a blind rule is to choose a non-negative integer Y from some specific distribution, and stop after Y steps.

The simplest blind stopping rule is the stopping rule used most often: "stop after t steps." Several other practical methods to generate elements from the stationary distribution (approximately) can also be viewed as blind rules. Stopping a lazy or continuous time random walk after a fixed number of steps corresponds to a blind rule for the original (discrete time) walk.

Our results above say that if we only want approximate mixing, then blind rules do essentially as well as any more sophisticated rule. The situation is very different if we want *exact* sampling. One cannot generate any distribution by a blind stopping rule; for example, starting from the stationary distribution, every blind rule generates the stationary distribution itself. We shall restrict our attention to stopping rules generating the stationary distribution (or at least approximations of it). Even this distribution cannot always be generated by a blind rule. The next theorem gives a characterization for the existence of a blind stopping rule for the stationary distribution.

Theorem 5.6 *Let $\lambda_1, \ldots, \lambda_n$ be the eigenvalues of M, $\lambda_1 = 1$.*

(a) If λ_k is positive real for some $k \geq 2$, then there exists a node s from which no blind stopping rule can generate π.

(b) If every λ_k, $k \geq 2$, is either non-real, negative or zero, then from any starting distribution σ there is a finite blind stopping rule that generates π.

Interestingly, the condition formulated in the theorem is most restrictive for undirected graphs; then all the eigenvalues are real, and typically many of them are positive. If there are no multiple edges, only complete multipartite graphs give a spectrum with just one positive eigenvalue.

Almost blind rules for an unknown graph. Suppose that we do a random walk on a digraph that we do not know. We are told the number of vertices, and we are able to recognize a node if we have seen it before. It is easy to argue that no matter how long we observe the walk, it is impossible to compute the stationary distribution exactly. Thus it is a bit surprising that one can *achieve* it exactly. Nonetheless that is what is done by Asmussen, Glynn and Thorisson [9]: they give a stopping rule where the probability of stopping after a walk $w_0 w_1 w_2 \dots$ depends only on the repetition pattern of nodes, and which produces a node from *exactly* the stationary distribution. The algorithm employed is complex and the expected number of steps required appears to be super-polynomial in the maximum hitting time $H(G)$, although no bound or estimate is given in the paper.

Note the emphasis on "exactly". If we only require that the last node be approximately from the stationary distribution, then a natural thing to do is to stop, say, after Y steps, where Y is chosen, say, randomly and uniformly from a sufficiently long interval. It is not at all obvious how to know (just by observing the walk) how long is "sufficiently long". But Aldous [7] describes a way to do so, and comes within total variation ε of the stationary distribution in time polynomial in $1/\varepsilon$ and linear in the maximum hitting time of the graph.

In [39] we describe a simple stopping rule which can reach the stationary distribution exactly, in any strongly connected digraph G. The rule requires only coin-flips for its randomization and can even be made deterministic unless the digraph is a single cycle (possibly with multiple edges). The expected number of steps is bounded by a polynomial in the maximum hitting time $H(G)$ of the graph.

The idea of the construction is to use formula (2) for the stationary distribution. Choose a node v *uniformly* from the set of all nodes. While observing the walk, mark the first exit from each node other than v. The edges we mark can be viewed as independent choices of one edge out of each node different from v. Hence given $v = i$, the probability that the $n - 1$ edges we marked form a spanning tree is $A_i \big/ \prod_{j \neq i} d_j$ So by (2), the probability of getting an in-arborescence rooted at i, conditional on getting an arborescence at all, is just π_i.

Thus if the edges of first exits form an arborescence, we can walk until we hit v and stop; else, we start again.

Unfortunately, the probability of getting an arborescence may be exponentially small, which would result in an exponentially long algorithm (in expected time). The trick is to replace the digraph by one whose adjacency matrix is a sufficiently large power of $I + A$; we omit the details here.

Other mixing measures. Theorems 5.1 and 5.2, and, in a weaker way, 5.3 are special cases of a surprising phenomenon, first explored by Aldous ([5], [6]). Mixing parameters of a random walk, that are only loosely related by their definition, are often very close. In fact, there seem to be three groups of parameters; within each group, any two are within (reasonably small) absolute constant factors to each other. For the time-reversible case (where these results are due to Aldous), the number of groups reduces to 2. We give a little "random walk" through some of these mixing measures.

Hitting times to sets. Let S denote a set of nodes and let $H(i, S)$ denote the expected number of steps before a random walk starting at i hits the set S. Of course, this number is larger if S is smaller, so it makes sense to scale by the stationary probability of S and define $T_{set} = \max_{s \in V, S \subseteq V} \pi(S) H(s, S)$. The upper bound in the following theorem is (in a somewhat different setting) due to Aldous, who also proved the lower bound for undirected graphs. The lower bound follows by an analysis of the shopping rule.

Theorem 5.7

$$\frac{1}{10} T_{tv} \leq T_{set} \leq 5 T_{tv}.$$

We remark that sometimes the following upper bound may be stronger:

$$T_{set} \leq T_{mix}$$

(but here a reverse inequality can only be claimed in the undirected case).

Forget time and reset time. From the point of view of statistics, the following measure of mixing is important. The "forget time" T_{forget} of a random walk is defined as the minimum mean length of any stopping rule that yields a distribution τ from the worst-case starting distribution for τ. In other words,

$$T_{forget} = \min_{\tau} \max_{\sigma} H(\sigma, \tau) = \min_{\tau} \max_{s} H(s, \tau)$$

(since the worst starting distribution for any given target distribution τ is clearly concentrated on a single node). This notion is central to the modern theory of Harris-recurrent chains; see e.g. [8].

In applications to sampling algorithms, we almost always have to draw repeated samples; are later samples cheaper than the first sample? More exactly, suppose that we have a node j from the stationary distribution; how long do we have to walk to generate another node, also from the stationary distribution, independent of the first? It is clear that the optimum stopping rule for this task is to follow an optimal stopping rule from j to the stationary distribution; so this stopping rule has mean length

$$T_{reset} = \sum_j \pi_j H(j, \pi),$$

which we call the *reset time* of the random walk. Trivially, $T_{\text{reset}} \leq T_{\text{mix}}$. The following result is proved in [40].

Theorem 5.8 *If the graph is undirected, then* $T_{\text{forget}} = T_{\text{reset}}$.

In the case of directed graphs, these two values may be arbitrarily far apart. But the theorem can be generalized to arbitrary digraphs in the form of an explicit formula for the forget time:

Theorem 5.9 *For every digraph,*

$$T_{\text{forget}} = \sum_j \pi_j \max_i H(i,j) - \sum_j \pi_j H(\pi, j).$$

($H(\pi, j)$ on the right hand side could be replaced by $H(k, j)$ with any $k \in V$ by (5).)

Using this formula, one can prove the following inequalities (for the case of undirected graphs, they were proved by Aldous [5]).

Theorem 5.10 *For every digraph,*

$$T_{\text{set}} \leq T_{\text{forget}} \leq 6T_{\text{tv}}.$$

(Hence $(1/10)T_{\text{tv}} \leq T_{\text{forget}} \leq 6T_{\text{tv}}$.) We conjecture that there is a constant c such that for any digraph, $T_{\text{mix}} \leq cT_{\text{reset}}$.

Maximum time and pointwise mixing. We have seen that the threshold rule was also optimal from the point of view that it minimizes the *maximum* number of steps needed to achieve the target distribution. If the target distribution is the stationary distribution, then we denote this maximum by T_{max}. This value may be quite different from the mean length of optimal stopping rules, even for undirected graphs. For example, let G be K_2 with N loops added on one of the nodes a single loop added on the other. It is easy to compute that $T_{\text{mix}} = (2N + 2)/(N + 3) \approx 2$, while (starting from the node with one loop), we need about $\log N$ steps to decrease the probability of staying there to the stationary value $2/(N + 3)$. Thus $T_{\text{max}} \approx \log N$.

A little unexpectedly, this value is also tied to mixing properties of the walk. Suppose that we want to generate any distribution τ such that $(9/10)\pi_i \leq \tau_i \leq (10/9)\pi_i$. If we allow an arbitrary stopping rule, then the time needed for this is clearly between T_{fill} and T_{mix}, and since these two values are close by Theorem 5.1, we don't get anything new.

However, the situation changes if we use a *blind rule*. Let T_{pw} denote the smallest T such that there exists a blind rule with maximum length T that produces (from every starting distribution) a distribution τ such that $(9/10)\pi_i \leq \tau_i \leq (10/9)\pi_i$.

Theorem 5.11

$$T_{\max} \leq 2T_{\mathrm{pw}}$$

In particular, it follows that if we take, say, the uniform averaging rule then we have to average over the first $\Omega(T_{\max})$ steps to achieve pointwise mixing (while in the filling sense, we only need $O(T_{\mathrm{mix}})$ steps, and to achieve mixing in total variation distance, we only need $O(T_{\mathrm{forget}})$ steps.

We conjecture that a reverse inequality also holds, in fact, averaging over $O(T_{\max})$ steps yields a distribution that is pointwise close to the stationary.

6   Chip-firing

Let G be a strongly connected directed graph (many of the results below extend to general digraphs, but for simplicity of presentation we restrict our attention to the strongly connected case). Let us place a pile of s_i chips on each node i of G. Recall that *firing a node* means to move one chip from this node to each of its children. Clearly a node can be fired iff $s_i \geq d_i^+$. If no node can be fired, we call the vector $s = (s_i)$ a *terminal configuration*. A (finite or infinite) sequence of firings is called a *chip-firing game*. The sequence of points fired is called the *record* of the game. We denote by $\mathcal{L}_s$ the set of of all records of finite games starting from the same initial configuration s, where the digraph $G = (V, E)$ is fixed. For $\alpha \in \mathcal{L}_s$, we denote by $|\alpha|$ the length of α. The multiset of nodes occuring in α is called the *score* of α.

The following properties of $\mathcal{L}$ have been proved in [13] for the undirected case, and extended to the directed case in [14]; they are also closely related to properties of abelian sandpiles proved by Dhar [20].

Proposition 6.1 *The set $\mathcal{L}_s$ of records of chip-firing games starting with the same configuration s has the following properties:*

(a) *Left-hereditary: whenever it contains a string, it contains all initial segments of the string.*

(b) *Permutable: whenever $\alpha, \beta \in \mathcal{L}_s$ have the same score, and $\alpha x \in \mathcal{L}_s$ for some $x \in V$, we also have $\beta x \in \mathcal{L}_s$.*

(c) *Locally free: whenever $\alpha x \in \mathcal{L}_s$ and $\alpha y \in \mathcal{L}_s$ for two distinct nodes $x, y \in V$, we also have $\alpha x y \in \mathcal{L}_s$.*

It turns out that these three simple properties have rather strong consequences. For example, it implies the following "antimatroid exchange property" (cf Korte, Lovász and Schrader [34]):

Proposition 6.2 *If $\alpha, \beta \in \mathcal{L}$ then there exists a subword α' of α such that $\beta\alpha' \in \mathcal{L}$ and the multiplicity of any v in $\beta\alpha'$ is the maximum of its multiplicities in α and β.*

The following theorem summarizes some of the results from [13], obtained using the above properties. It asserts that chip-firing games have a certain "Church-Rosser" property.

Theorem 6.3 *For a given directed graph G and initial distribution s of chips, either every chip-firing game can be continued indefinitely, or every game terminates after the same number of moves with the same terminal position. The number of times a given node is fired is the same in every terminating game. If a game is infinite, then every node gets fired infinitely often.*

In the case of undirected graphs, Tardos [48] proved a strong converse of the last assertion:

Lemma 6.4 *If a chip-firing game on an undirected graph is finite, then there is a node that is never fired.*

This assertion is analogous to Theorem 4.5 for stopping rules; however, it does not remain true for general digraphs. It was shown in [14] that it remains true for eulerian digraphs, and that it can be extended to digraphs in a different way (see Lemma 6.6 below).

Given a graph, we may ask: what is the minimum number of chips that allows an infinite game? What is the maximum number of chips that allows a finite game? In [13] it was shown that for an undirected graph with n nodes and m edges, more than $2m - n$ chips guarantees that the game is infinite; fewer than m chips guarantee that the game is finite; for every number N of chips with $m \leq N \leq 2m - n$, there are initial positions that lead to an infinite game and initial positions that lead to a finite game.

For directed graphs, the second question can still be answered trivially: if G is a directed graph with n nodes and m edges, and we have $N > m - n$ chips, then the game is infinite (there is always a node that can be fired, by the pigeonhole principle), and $N \leq n - m$ chips can be placed so that the game terminates in 0 steps.

It is not known how to determine the minimum number of chips allowing an infinite game on a general digraph. This is not just a function of the number of nodes and edges. For eulerian digraphs, it was mentioned in a remark added in proof to [14] that the minimum number of chips that can start an infinite game is the *edge-feedback number*, i.e., the minimum number of edges whose removal destroys all directed cycles. Moreover, the feedback number is always a lower bound on the number of chips in an infinite game.

Chip conservation. A useful tool in the study of chip-firing games is the following "chip conservation equation" from [13] (cf. Lemma 4.1). Let s be the initial and t, the final configuration of a finite game, and let x_i denote the number of times the node i is fired. Let a_{ij} be the number of edges from node i to node j. Then

$$\sum_j a_{ji} x_j - d_i^+ x_i = t_i - s_i. \tag{13}$$

Let $L \in \mathbb{R}^{V \times V}$ be the matrix defined by

$$L_{ij} = \begin{cases} a_{ij}, & \text{if } i \neq j, \\ -d_i^+, & \text{if } i = j. \end{cases}$$

We call L the *Laplacian* of the digraph G. Note that $L\mathbf{1} = 0$, so L is singular. It is also well known that for strongly connected digraphs, the co-rank of L is 1. Let $v = (v_i)$ denote the solution of $L^T v = 0$, scaled so that the v_i are coprime integers. From the Frobenius-Perron theory it follows that we may assume that $v > 0$. If G is an digraph (in particular, if G is an undirected graph), then $v = \mathbf{1}$. The quantities $|v| := \sum_i v_i$ and $\|v\| := \sum_i d_i v_i$ play an important role in chip-firing.

The Laplacian is also related to the transition probability matrix of the random walk:

$$M = D^{-1}L + I,$$

where D is the diagonal matrix with the outdegrees in the diagonal. It follows that the stationary probabilities are given by

$$\pi_i = \frac{d_i^+ v_i}{\|v\|}.$$

It follows by (3) that the maximum hitting time can be estimated as follows:

$$H(G) \leq \|v\| \sum_i \frac{1}{v_i} \leq n\|v\|. \tag{14}$$

In terms of the Laplacian, equation (13) can be written as

$$L^T x = t - s.$$

Period length. As a first application of this identity, we discuss periodic games. More exactly, consider a period, i.e., a game that starts and ends with the same configuration. Let x be its score vector; then

$$L^T x = 0,$$

whence it follows that $x = tv$ for some positive integer t. It is easy to see that $x = v$ can be realized: just place a very large number of chips on each node, and fire each node i v_i times in any order. The conservation equation implies that we return to the starting configuration.

A key property of the vector v is the following:

Lemma 6.5 *Let $\alpha \in \mathcal{L}_s$. Let α' be obtained by deleting the first v_i occurrences of node i in α (if i occurs fewer then v_i times, we delete all of its occurrences). Then α' is the record of a game from the same initial position.*

This lemma (which is easy to prove by counting chips) has a number of consequences. First, 6.2 implies that the deleted elements can be added to α', and so we get a game that is a rearrangement of α but starts with α'. From this it is easy to derive that *if a configuration starts a periodic game, it also starts one with period score v.* We also get an extension of Lemma 6.4:

Lemma 6.6 *In every terminating game, there is a node i that is fired fewer than v_i times.*

From these considerations, one obtains:

Proposition 6.7 *The minimum length of a period of any game on the graph G is $|v|$, and the number of chips moved during a minimal period is $\|v\|$.*

Game length. Deviating from earlier papers, we measure the length of a game by the number of chip-motions (so the firing of a node of outdegree d contributes d to the length). This is of course an upper bound on the number of firings, and is never more than a factor of m larger.

Tardos [48] proved that on an undirected graph, every terminating game ends in a polynomial number of steps. We sketch a new proof based on the conservation equation. Consider a game that terminates, and let z_i be the number of times node i is fired. Then we have

$$\sum_j a_{ji} z_j - d_i z_i = t_i - s_i.$$

Here, by termination, $s_i < d_i$. We can rewrite this equation as

$$\sum_j \frac{a_{ij}}{d_j} \cdot \frac{d_j z_j}{m} - \frac{d_i z_i}{m} = \frac{t_i + d_i - s_i}{m} - \frac{d_i}{m}.$$

Thus the numbers $d_i z_i / m$ are the exit frequencies of a stopping rule from the distribution defined by $\tau_i = (t_i + d_i - s_i)/m$ to π. By Lemma 6.4, the minimum of these exit frequencies is 0, and hence the stopping rule is optimal. This implies that

$$\sum_i \frac{d_i z_i}{m} = H(\tau, \pi) \le T_{\mathrm{mix}}.$$

Hence we get:

Theorem 6.8 *The number of chips moved during a terminating game on an undirected graph G is at most $m T_{\mathrm{mix}}$.*

Eriksson [28] showed that on a directed graph (even on a graph with all but one edges undirected) a terminating game can be exponentially long. It was proved in [14] that the maximum length of a terminating game can exceed the period length by a polynomial factor only. It was conjectured that a converse inequality, bounding the period length by a polynomial multiple of the maximum game length, also holds. It turns out that this conjecture is true, and in fact it follows quite simply using the conservation equation. Results on random walks discussed above also yield an improvement in the first direction.

Theorem 6.9 *Let M denote the maximum number of chip-motions in a terminating finite game. Then*

$$\|v\| - m \le M \le nm\|v\|.$$

Sketch of proof. 1. Consider a terminating game, and let p and q be the beginning and terminating configurations of it. Obviously, $|p| = |q| < m$. Let u be the score vector of the game. By Lemma 6.6, there is a node i such that $u_i < v_i$. Hence we can write $u = tv + w$, where $0 \le t < 1$ and $\min w_i = 0$. By the conservation equation, we have

$$L^{\mathsf{T}} u = q - p,$$

and hence we also have

$$L^{\mathsf{T}} w = q - p.$$

Let N denote the number of chips, then $N \le m - n$ since the game terminates. We get that the numbers $x_i = d_i^+ w_j / N$ are the exit frequencies of an optimum stopping rule from $(1/N)p$ to $(1/N)q$. This implies that

$$\sum_i \frac{d_j^+ w_j}{N} = H\left(\frac{1}{N}p, \frac{1}{N}q\right) \le H(G).$$

Thus by (14), the number of chips moved is

$$\sum_i d_i^+ u_i = t \sum_i d_i^+ v_i + \sum_i d_i^+ w_i < \|v\| + N H(G) < nm\|v\|.$$

2. To prove the other inequality (and thereby verify a conjecture from [14]), place $d_i^- - 1$ chips on node i. We claim that every game from this starting position is finite; in fact, we claim that no node can be fired v_i times. Assume that this is false, and consider the first step when a node i is fired the v_i-th time. Let y be the score vector up to and including this step, and q, the configuration after this step. Then the conservation equation says:

$$\sum_j a_{ji} y_j - d_i^+ y_i = q_i - (d_i^- - 1).$$

But here the left hand side is

$$\sum_j a_{ji} y_j - d_i^+ y_i \leq \sum_j a_{ji}(v_j - 1) + d_i^+ v_i = -\sum_j a_{ji} = -d_i^-,$$

which is a contradiction since $q_i \geq 0$.

Thus every game started from this position terminates. But with $m - n$ chips on board, the only terminating configuration is having $d_i^+ - 1$ chips on node i. Moreover, substitution into the chip conservation equation shows that in order to get from $d_i^- - 1$ chips to $d_i^+ - 1$ chips on each node, we have to fire every node i exactly $v_i - 1$ times. Hence there is always a terminating game of length

$$\sum_i d_i(v_i - 1) = \|v\| - m.$$

$\square$

We have seen three relations between the three diffusion parameters we considered: the maximum hitting time $H(G)$, the period length $\|v\|$, and the maximum game length M. The last two are equal up to a polynomial factor, while the first is at most this large.

The hitting time can be much smaller than the other two quantities. Consider a 2-connected undirected graph G and orient one edge (leave the rest two-way). Then one can argue that the hitting time remains polynomial; on the other hand, the example of Eriksson mentioned above is of this type, and here the game length and period length are exponentially large.

Algorithmic issues. Results mentioned above were used in [14] to give an algorithm for checking whether a given position on an undirected graph can be transformed to another given position by a sequence of firings. The running time of the algorithm is polynomial in the period length $\|v\|$, so in the case of undirected graphs, it is polynomial in m. The idea is to use Lemma 6.6 in a manner similar to the proof of Theorem 6.9 to show that if there is a sequence of chip-firings then there is one of length polynomial in $\|v\|$, and in fact the firing frequencies z_i can be calculated by simple arithmetic. Then one can show that any game with the additional restriction that no node i is fired more than z_i times, must terminate in the prescribed target position, or else the target position is not reachable.

Unfortunately, no truly polynomial algorithm is known to decide the reachability question. It is also not known how to decide in polynomial time whether a given initial position starts a finite or infinite game.

These questions are quite interesting because chip-firing on a digraph may be considered as a "totally asynchronous" distributed protocol (by Theorem 6.3). The comparison of the class of functions computable by such a protocol with the class P seems both interesting and difficult.

Avalanches. Let each node of a digraph represent a site where snow is accumulating. One special node s is considered the "outside world". Once the amount of snow on a site (other than s) surpasses a given threshold, the site can "break", sending one unit of snow to each of its out-neighbors. This may result in overloading some of the children of the node, and then these nodes break etc. If the digraph is strongly connected (which we assume for simplicity) then after a finite number of steps, no node will have too much snow (except s, which cannot break), and the avalanche terminates.

To maintain the dynamics of the model, snow is supposed to fall on the nodes. There are various ways to model this; simplest of these is to assume that each node i gets a given a_i amount of snow in unit time. We add snow until some node reaches the breaking threshold and starts an avalanche again (which happens so fast that no new snow falls during the avalanche).

The breaking threshold can be chosen, after some easy reductions, to be the outdegree of the node; then the avalanche is just a chip-firing game (where s is not allowed to be fired). But we can also include snow-fall in this model: we connect s to node i by a_i edges. Then a snowfall just corresponds to firing node s. We assume that there is enough snow in s (all those oceans, snow-caps etc) so that it can always be fired.

Thus a sequence of avalanche–snowfall–avalanche–snowfall–... is just an infinite chip-firing game on the graph, with the additional restriction that the special node s is only fired if no other node can be fired. We call such a restricted chip-firing game an *avalanche game*. When an avalanche starts, it consists of a sequence of firings which may happen in many ways, but the length of the avalanche, the number of times a node is fired, as well as the ending position are uniquely determined. The ending position of an avalanche is called *stable*.

Consider a periodic avalanche game. The (stable) position immediately before snowfall is called a *recurrent* position. A snowfall followed by an avalanche leads to another recurrent position, and this defines a permutation of recurrent positions. Each cycle in this permutation corresponds to a periodic avalanche game. The score vector of this game is an integer multiple of the period vector v. It follows by an argument almost identical to the second half of the proof of Theorem 6.9 that in fact we get the period vector. Hence the number of recurrent positions in the cycle is v_s. It follows that the average length of an avalanche is

$$\frac{1}{v_s} \sum_{i \neq s} v_i,$$

independently of the cycle.

The conservation equation is very useful in the study of recurrent configurations. Gabrielov ([31]) introduces the lattice

$$\mathcal{L} = \{L^\mathsf{T} u : u \in \mathbb{Z}^V, u_s = 0\}$$

The conservation equation implies that if a position p can be obtained from a position q by a sequence of firings of nodes different from s, then $p - q \in \mathcal{L}$. It is not difficult to prove that if two positions p and q satisfy $p - q \in \mathcal{L}$ and p_i, q_i are large enough for all $i \neq s$, then there is a position that can be reached from each of p and q. Hence the stable positions in which the avalanches starting from p and q end are the same. It is also easy to see that this position is recurrent. These considerations imply that every translated copy $u + \mathcal{L}$ of the lattice $\mathcal{L}$ (with $u \in \mathbb{Z}^V$) contains a unique recurrent position. Thus the number of recurrent positions is the same as the number of cosets of $\mathcal{L}$ in $\mathbb{Z}^V$, which is $\det(\mathcal{L})$. Hence an easy computation gives the following interesting theorem of Dhar ([20]):

Theorem 6.10 *The number of recurrent positions is $\det(L')$, where L' is the matrix obtained from L by deleting the row and column corresponding to the node s.*

The reader may recognize that this determinant is just the number of spanning arborescences of G rooted at s, by the "Matrix-Tree Theorem" of Tutte. This relation is explained and exploited in [30], [31].

There are many characterizations of recurrent positions. For example, a position p is recurrent if and only if there exists a position q with $p_i \leq q_i$ for each node $i \neq s$ such that the avalanche starting from q ends with p.

Speer [46] gives a characterization that gives a way to test for recurrence. To describe this, we introduce a version of the period vector. We say that a vector $v \in \mathbb{Z}_+^V$ is *reducing*, if $v_s = 0$ and starting with N chips on each node (where N is a large integer), and firing each node i v_i times, we obtain a position with at most N chips on each node $i \neq s$. It is easy to see that a reducing vector must satisfy $v_i > 0$ for $i \neq s$. So we may fire every node once right away. This may produce a position with more than N chips on some node; this node must be fired at least twice during the game, so we may as well fire it right away, etc. This way we construct a "canonical" reducing vector $\hat{v}$ such that $\hat{v} \leq v$ for every reducing vector v.

Now one can prove that following analogue of Lemma 6.5:

Lemma 6.11 *Let $\alpha \in \mathcal{L}_p$, and assume that α does not contain s. Let α' be obtained by deleting the first $\hat{v}_i$ occurrences of node i from α (if i occurs fewer then v_i times, we delete all of its occurrences). Then $\alpha' \in \mathcal{L}_p$.*

Corollary 6.12 *A stable configuration p is recurrent if and only if the avalanche starting from $p - L^\mathsf{T}\hat{v}$ ends with p.*

References

[1] R. Aleliunas, R.M. Karp, R.J. Lipton, L. Lovász and C.W. Rackoff, Random walks, universal travelling sequences, and the complexity of maze problems, *Proc. 20th Ann. Symp. on Foundations of Computer Science* (1979), 218–223.

[2] W. Aiello, B. Awerbuch, B. Maggs and S. Rao, Approximate load balancing on dynamic and asynchronous networks, *Proc. 25th ACM Symp. of Theory of Computing* (1993), 632–634.

[3] N. Alon, Eigenvalues and expanders, *Combinatorica* **6** (1986), 83–96.

[4] R.J. Anderson, L. Lovász, P.W. Shor, J. Spencer, É. Tardos and S. Winograd, Disks, balls, and walls: analysis of a combinatorial game, *Amer. Math. Monthly* **96** (1989), 481–493.

[5] D.J. Aldous, *Reversible Markov Chains and Random Walks on Graphs* (book), to appear.

[6] D.J. Aldous, Some inequalities for reversible Markov chains, *J. London Math. Soc.* **25** (1982), 564–576.

[7] D.J. Aldous, On simulating a Markov chain stationary distribution when transition probabilities are unknown, preprint (1993).

[8] S. Asmussen, *Applied Probability and Queues*, Wiley, New York 1987.

[9] S. Asmussen, P. W. Glynn and H. Thorisson, Stationary detection in the initial transient problem, *ACM Transactions on Modeling and Computer Simulation* **2** (1992), 130–157.

[10] J.R. Baxter and R.V. Chacon, Stopping times for recurrent Markov processes, *Illinois J. Math.* **20** (1976), 467–475.

[11] D. Bayer and P. Diaconis, Trailing the dovetail shuffle to its lair, *Ann. Appl. Probab.* **2** (1992), 294–313.

[12] P. Bak, C. Tang and K. Wiesenfeld, Self-organized criticality, *Physical Revue A* **38** (1988), 364–374.

[13] A. Björner, L. Lovász and P. Shor, Chip-firing games on graphs, *Europ. J. Comb.* **12** (1991), 283–291.

[14] A. Björner and L. Lovász, Chip-firing games on directed graphs, *J. Algebraic Combinatorics* **1** (1992), 305 328.

[15] G. Brightwell and P. Winkler, Maximum hitting time for random walks on graphs, *J. Random Structures and Algorithms* **1** (1990), 263–276.

[16] R.V. Chacon and D.S. Ornstein, A general ergodic theorem, *Illinois J. Math.* **4** (1960), 153–160.

[17] F.R.K. Chung and S.-T. Yau, Eigenvalues of graphs and Sobolev inequalities, to appear.

[18] A.K. Chandra, P. Raghavan, W.L. Ruzzo, R. Smolensky and P. Tiwari, The electrical resistance of a graph captures its commute and cover times, *Proc. 21st ACM Annual ACM Symposium on the Theory of Computing* (1989), 574–586.

[19] D. Coppersmith, P. Tetali and P. Winkler, Collisions among Random Walks on a Graph, *SIAM J. on Discrete Mathematics* **6** No. 3 (1993), 363–374.

[20] D. Dhar, Self-organized critical state of sandpile automaton models, *Physical Revue Letters* **64** (1990), 1613–1616.

[21] P. Diaconis, *Group Representations in Probability and Statistics*, Inst. of Math. Statistics, Hayward, CA (1988).

[22] P. Diaconis and D. Stroock, Geometric bounds for eigenvalues of Markov chains, *Annals of Appl. Prob.* **1** (1991), 36–62.

[23] J. Dodziuk and W.S. Kendall, Combinatorial Laplacians and isoperimetric inequality, in: *From Local Times to Global Geometry, Control and Physics*, (ed. K. D. Ellworthy), Pitman Res. Notes in Math. Series **150** (1986), 68–74.

[24] P.G. Doyle and J.L. Snell, *Random Walks and Electric Networks*, Mathematical Assoc. of America, Washington, DC 1984.

[25] M. Dyer, A. Frieze and R. Kannan, A random polynomial time algorithm for estimating volumes of convex bodies, *Proc. 21st Annual ACM Symposium on the Theory of Computing* (1989), 375–381.

[26] A. Engel, The probabilistic abacus, *Educ. Stud. in Math.* **6** (1975), 1–22.

[27] A. Engel, Why does the probabilistic abacus work? *Educ. Stud. in Math.* **7** (1976), 59–69.

[28] K. Eriksson, No polynomial bound for the chip-firing game on directed graphs, *Proc. Amer. Math. Soc.* **112** (1991), 1203–1205.

[29] J.A. Fill, Eigenvalue bounds on convergence to stationary for nonreversible Markov chains, with an application to the exclusion process, *Ann. of Appl. Prob.* **1** (1991), 62–87.

[30] A. Gabrielov, Avalanches, sandpiles, and Tutte decomposition for directed graphs, preprint (1993).

[31] A. Gabrielov, Asymmetric abelian avalanches and sandpiles, preprint (1993).

[32] B. Ghosh and S. Muthukrishnan, Dynamic load balancing by random matchings, *J. Comp. Sys. Sci*, to appear.

[33] M.R. Jerrum and A. Sinclair, Approximating the permanent, *SIAM J. Comput.* **18** (1989), 1149–1178.

[34] B. Korte, L. Lovász and R. Schrader, *Greedoids*, Springer, 1991.

[35] L. Lovász, Random walks on graphs: a survey, in: *Combinatorics, Paul Erdős is Eighty*, (eds. D. Miklós, V.T.Sós and T. Szőnyi) J. Bolyai Math. Soc., Vol. II, to appear.

[36] L. Lovász and M. Simonovits, Random walks in a convex body and an improved volume algorithm, *Random Structures and Alg.* **4** (1993), 359–412.

[37] L. Lovász and P. Winkler, A note on the last new vertex visited by a random walk, *J. Graph Theory* **17** (1993), 593–596.

[38] L. Lovász and P. Winkler, Exact mixing in an unknown Markov chain, preprint (1994).

[39] L. Lovász and P. Winkler, Fast mixing in a Markov chain, preprint (1994).

[40] L. Lovász and P. Winkler, The forget time of a Markov chain, preprint (1994).

[41] M. Mihail, Conductance and convergence of Markov chains: a combinatorial treatment of expanders, *Proc. 30th Ann. Symp. on Foundations of Computer Science* (1989), 526–531.

[42] J.W. Pitman, Occupation measures for Markov chains, *Adv. Appl. Prob.* **9** (1977), 69-86.

[43] W. Reisig, *Petri Nets: An Introduction*, Springer-Verlag, New York 1985.

[44] A. Skorokhod, *Studies in the Theory of Random Processes*, orig. pub. Addison-Wesley (1965), 2nd ed. Dover, New York 1982.

[45] A. Sinclair, Improved bounds for mixing rates of Markov chains and multicommodity flow, *Combinatorics, Probability and Computing* **1** (1992), 351–370.

[46] E. R. Speer, Asymmetric Abelian sandpile models, *J. Stat. Phys.* **71** (1993), 61–74.

[47] J. Spencer, Balancing vectors in the max norm, *Combinatorica* **6** (1986), 55–66.

[48] G. Tardos, Polynomial bound for a chip firing game on graphs *SIAM J. Disc. Math* **1** (1988), 397–398.

Cayley Graphs: Eigenvalues, Expanders and Random Walks

Alexander Lubotzky*

Institute of Mathematics

Hebrew University

Jerusalem 91904 ISRAEL

alexlub@math.huji.ac.il

INTRODUCTION

(0.1)   Let $\{G_i\}_{i \in I}$ be a family of finite groups and for each $i \in I$, let Σ_i be a symmetric (i.e., $\Sigma_i = \Sigma_i^{-1}$) set of generators for G_i. Let $X(G_i; \Sigma_i)$ be the Cayley graph of G_i with respect to Σ_i (i.e., the graphs whose vertices are the elements of G_i and $a \in G_i$ is adjacent to σa, $\sigma \in \Sigma_i$).

We will consider the following two questions:

(I) When does the family of graphs $X_i = X(G_i; \Sigma_i)$, $i \in I$, form a family of expanders? (This means that there exists a fixed $C > 0$ such that for every i and every $A \subseteq X_i$, with $|A| \leq \frac{1}{2}|X_i|$, $|\partial A| \geq C|A|$, when ∂A is the set of vertices of distance 1 from A).

(II) How fast the random walk on $X(G_i; \Sigma_i)$ is converging to the uniform distribution?

(0.2)   The basic problem we are interested in is to what extent the answers to questions I and II depend on the algebraic structure of the groups G_i and the choice of generators Σ_i.

The above questions can be studied in several different contexts:

---

*Partially sponsored by the Edmund Landau Center for research in Mathematical Analysis, supported by the Minerva Foundation (Germany).

(a) $\{|\Sigma_i| \mid i \in I\}$ is bounded or not.

(b) The Σ_i-s are chosen as "best case generators", "worst case generators" or "average (=random) generators".

Questions I and II are closely related to each other and both are connected with estimating the eigenvalues of the adjacency matrix of X_i.

(0.3)   This survey paper is organized as follows: In §1 we set notations and recall the basic results on eigenvalues of graphs. Beside the standard results we take the opportunity to report on some of the work of Greenberg [Gr] which deals with a more general notion of Ramanujan graphs. His work suggests some interesting problems related to groups and eigenvalues.

In §2, we treat Cayley graphs with a bounded number of generators. This is the most interesting and most difficult context. Some deep mathematical tools - such as: Kazhdan property (T), Selberg Theorem $\lambda_1 \geq 3/16$ and Ramanujan conjecture - have been used to construct families of expanders as Cayley graphs of some finite groups with respect to carefully chosen systems of generators of bounded cardinality (cf. [Lu], [LW] and the references therein). But changing the groups or even slight changes in the generators paralyze these deep theories. We survey this in §2 - but only briefly - referring the reader to a more detailed discussion in [LW].

The case when $|\Sigma_i|$ are not bounded is discussed in §3. Here also the theory is far from being satisfying - but some progress has been made recently, especially, for random generators or when Σ is a full conjugacy class. Alon and Roichman [AR] showed that $O(\log|G_i|)$ random generators give rise to expanders. This happens independently of the structure of the group G_i. This puts the "boundary conditions" and opens up the question for special type of groups (e.g., simple groups and the symmetric groups). Some sporadic results have been proved in this direction.

In §4 we turn to random walks. Question I about expanders is closely related to estimating the second largest eigenvalue. For random walks all eigenvalues are relevant. Random walks on groups is a rich subject (for an excellent treatment see [D]) of great importance in probability and statistics (e.g., for card shuffling problems). We have made no attempt to overview it. We merely concentrate on one corner: random walks on a simple group G when Σ is a full conjugacy class. Diaconis and Shahshahani [DS] have shown how to use representation theory of finite groups to tackle such a case and solved the case of $G_n = S_n$ and $\Sigma_n = $ the set of all transpositions. Recently Roichman ([R2], [R3]) essentially solved the problem for almost all conjugacy classes of S_n. His work opens up interesting questions for other finite simple groups (see §4.12), questions which can shed light on the combinatorics of finite simple groups.

In §5 we change gears: here we take the opportunity to describe an application of groups to eigenvalues of graphs. We show how to construct isospectral (=cospectral) k-regular graphs, i.e., pairs of graphs with the same set of eigenvalues. The method is a small variant of the method of Sunada [Su] who constructed isospectral Riemannian manifolds. A comprehensive study has been carried out by Brooks [Br].

Further background on some of the topics discussed in this survey can be found in [Lu], where the reader is also referred to for unexplained notions.

Acknowledgment The author acknowledges with gratitude discussions with Bob Brooks, Shahar Mozes, Gil Kalai, Nati Linial, Yuval Roichman and Avi Wigderson which have been helpful to me in writing this survey.

§1. Eigenvalues

(1.1)   Let $X = (V, E)$ be a connected undirected graph where $\deg(x) \leq$

k for every vertex x of X, when $\deg(x)$ denotes the number of edges coming out of x. Let $L^2(X)$ be the space of functions f on X (i.e., on V-the set of vertices of X) with $\sum_{x \in V} |f(x)|^2 < \infty$ and $\delta : L^2(X) \to L^2(X)$ be the **adjacency operator**, i.e. $(\delta f)(x) = \sum_y \delta_{xy}(f(y))$ when δ_{xy} denotes the number of edges from x to $y \in V$.

Denote $\rho(X)$ the **spectral radius** of δ, i.e. $\rho(X) = \sup\{|\lambda| \mid \lambda \in$ spectrum of $\delta\}$. It is well known (cf. [Lu, Chapter 4] and the references therein) that $\rho(X) = \limsup a_n^{1/n}$ when a_n is the number of closed paths of length n on X starting from a fixed vertex x_0 on X. One can deduce from this that if Y_1 and Y_2 are two graphs and $\pi : Y_1 \to Y_2$ is a cover map (i.e., a surjective map which is a local isomorphism, namely, for every $y \in Y_1$, π induces an isomorphism from $st(y)$ to $st(\pi(y))$, where $st(y)$ denotes the set of vertices of distance at most one from y) then $\rho(Y_1) \leq \rho(Y_2)$. If π is a finite cover then $\rho(Y_1) = \rho(Y_2)$. A theorem of Leighton [Le] asserts that any two finite graphs Y_1 and Y_2 with the same simply connected covering tree have a common finite cover Y. One can now deduce:

(1.2) PROPOSITION (GREENBERG [Gr]). *let X be a connected locally finite graph and let $\Omega_f(X)$ be the family of finite graphs covered by X. Then for $Y_1, Y_2 \in \Omega_f(X)$, $\rho(Y_1) = \rho(Y_2)$. This common value will be denoted $\chi(X)$.*

(1.3) EXAMPLE: (a) let $X = X_k$ be the infinite k-regular tree. Then $\chi(X) = k$ while $\rho(X) = 2\sqrt{k-1}$ (see [Lu, Chapter 4]).

(b) let $X = X_{m,n}$ be the infinite bi-partite (m,n)-bi-regular graphs. Then $\chi(X) = \sqrt{mn}$ and $\rho(X) = \sqrt{m-1} + \sqrt{n-1}$.

(c) Let Γ be a group, Σ a symmetric set of k generators for Γ and $X = X(\Gamma; \Sigma)$ the Cayley graph of Γ with respect to Σ. Then $\chi(X) = k$, which is always the case for a k-regular graph. On the other hand $\rho(X)$ reflects some deeper properties of Γ, e.g., from results of Kesten [Ke] it follows:

(i) $\rho(X) = 2\sqrt{k-1}$ if and only if X is a k-regular tree, i.e.
$\Sigma = \{x_1, x_1^{-1}, \ldots, x_t, x_t^{-1}, y_1, \ldots, y_\ell\}$ where each y_i $(1 \leq i \leq \ell)$ is an element of order 2, $k = 2t + \ell$ and Γ is the free product of the cyclic groups
$\Gamma = \overset{t}{\underset{i=1}{*}} \langle x_i \rangle * \overset{\ell}{\underset{j=1}{*}} \langle y_j \rangle$.

(ii) $\rho(X) = k$ if and only if Γ is an amenable group (see [Lu, p.14] for a definition of amenable groups and their relation to expanders).

(iii) More generally, if N is a normal subgroup of Γ, then $\rho(X(\Gamma/N; \Sigma)) = \rho(X(\Gamma; \Sigma))$ if and only if N is an amenable group. ((ii) follows from (iii), since $\rho(Y) = k$ for the bouquet of k circles).

(iv) It follows from (iii) that if Γ is neither amenable nor a free product of cyclic groups, then $2\sqrt{k-1} < \rho(X(\Gamma; \Sigma)) < k$. This applies in particular to infinite groups with Kazhdan property (T).

(1.4)   Let X be a fixed, connected, infinite locally finite graph. For $Y \in \Omega_f(X)$ of order n, denote $\lambda_0(Y) > \lambda_1(Y) \geq \lambda_2(Y) \geq \ldots \geq \lambda_{n-1}(Y)$ the eigenvalues of Y and $\mathrm{spec}(Y) = \{\lambda_0(Y), \ldots, \lambda_{n-1}(Y)\}$. It follows from the Perron-Frobenious Theorem that $\lambda_0(Y) = \chi(X)$.

THEOREM (GREENBERG [Gr]). *Given $\varepsilon > 0$ there exists $c = c(X, \varepsilon)$, $0 < c < 1$, such that for every $Y \in \Omega_f(X)$, $|\{\lambda \in \mathrm{spec}(Y) \mid \lambda \leq \rho(X) - \varepsilon\}| < c|Y|$ and $|\{\lambda \in \mathrm{spec}(Y) | \lambda \geq -\rho(X) + \varepsilon\}| < c|Y|$, i.e., a fraction at least $(1 - c)$ of the eigenvalues of Y are greater than $\rho(X) - \varepsilon$, and at least $(1 - c)$ are smaller then $-\rho(X) + \varepsilon$.*

(1.5)   Theorem 1.4 is a far reaching generalization of the following well known result of Alon and Bopanna (It also extends some unpublished results of M. Burger and of J.P. Serre).

THEOREM. *If X_n be an infinite family of k-regular graphs (k fixed) then $\liminf \lambda_1(X_{n,k}) \geq 2\sqrt{k-1}$.*

(1.6)   Theorem 1.5 has been the motivation for the definition of Ramanujan graphs for k-regular graphs. A finite k-regular graph Y is called Ramanujan if for every eigenvalue λ of Y, either $\lambda = \pm k$ or $|\lambda| \leq 2\sqrt{k-1}$. Theorem 1.4 justifies the following:

DEFINITION: A finite graph Y covered by an infinite graph X will be called X-**Ramanujan** if for every eigenvalue λ of Y, either $\lambda = \pm\chi(X)$ or $|\lambda| \leq \rho(X)$. Y is a **Ramanujan graph** if it is $\tilde{Y}$-Ramanujan when $\tilde{Y}$ is the universal tree cover of Y.

(1.7)   Explicit constructions of k-regular Ramanujan graphs were given by Lubotzky, Phillips and Sarnak (cf. [Lu]) for various values of k. The most general result is due to Morgenstern (cf. [Lu, Chapter 8]) who constructed infinitely many for every k of type $p^{\alpha} + 1$ where p is an arbitrary prime. All these constructions have used deep number theory. It is quite unlikely that the existence of infinitely many k-regular Ramanujan graphs depends on k. Still the following is open:

PROBLEM: Prove that for every $k \geq 3$, there are infinitely many k-regular Ramanujan graphs.

(1.8)   It is less clear when a general infinite tree covers finite Ramanujan graphs. Of course, not every tree X covers a finite graph. By covering theory a necessary condition is that $\mathrm{Aut}(X)$ has only finitely many orbits on X. But this is not sufficient. A necessary and sufficient condition for covering a finite graph was given by Bass and Kulkarni [BK]. A tree which covers a finite graph is called a **uniform tree**.

PROBLEM: Does every infinite uniform tree cover a Ramanujan graph? Infinitely many such graphs?

(1.9)   Our main interest in the current paper is in Cayley graphs. Theorem 1.4 has some corollaries for those and open some interesting problems:

COROLLARY. *Assume* Γ *is a group which is not a free product of cyclic groups,* Σ *a finite symmetric set of* k *generators and* $\{N_i\}_{i \in I}$ *a family of finite index subgroups of* Γ. *Then* $X(\Gamma/N_i; \Sigma)$ *are Ramanujan for at most finitely many* $i \in I$.

PROOF: By (1.3c), $\rho(X(\Gamma; \Sigma)) > 2\sqrt{k-1}$, so use Theorem 1.4 with $\varepsilon = \frac{\rho(X(\Gamma;\Sigma))-2\sqrt{k-1}}{2}$.

Recall that the first method to construct expanders was using Cayley graph quotients of groups with property (T). (E.g.,

$$X\left(SL_3(\mathbb{F}_p); \left\{ \begin{pmatrix} 1 & 1 & 0 \\ 0 & 1 & 0 \\ 0 & 0 & 1 \end{pmatrix}^{\pm 1}, \begin{pmatrix} 0 & 1 & 0 \\ 0 & 0 & 1 \\ 1 & 0 & 0 \end{pmatrix}^{\pm 1} \right\} \right),$$

see (Lu, §3]). The Corollary implies that these are **not** Ramanujan.

(1.10)   So, given $X = X(\Gamma; \Sigma)$ as in (1.9) we can at best hope to have X-Ramanujan graphs in $\Omega_f(X)$. This is not always the case! From elementary covering theory it follows that finite graphs covered by X are Schreier graphs (see §5) corresponding to some finite index subgroups of Γ. It is well known that there are finitely generated groups with no proper finite index subgroups (cf. [Se, p. 9]). Thus for such groups Γ, X will not cover X-Ramanujan graphs (with the exception of the trivial bouquet of circles).

Recall that a group Γ is said to be **residually finite** if the intersection of its finite index subgroups is trivial. The following is an interesting open problem:

PROBLEM: Let Γ be a residually finite infinite group, Σ a symmetric set of generators and $X = X(\Gamma; \Sigma)$. Are there infinitely many X-Ramanujan graphs covered by X?

This Problem is an extension of Problem 1.7 (since a k-regular tree is Cayley graph of a residually finite group, e.g. the free product of k cyclic groups of order 2, or if k is even, the free group on $k/2$ generators). While it is very likely that problem 1.7 has an affirmative answer and probably also Problem 1.8, we do not know what to expect as an answer to Problem 1.10.

(1.11)   We close this section by recalling the well known connection between eigenvalues and the expansion of graphs.

DEFINITION: Let X be a finite graph. Denote by $C(X) = \inf \left\{ \frac{|\partial A|}{|A|} \mid A \subseteq X \right.$ with $|A| \le \frac{1}{2}|X| \}$. $C(X)$ is called the **expansion constant** of X. It is the largest constant C for which X is a C-expander (see (0.1)).

The following result has been proved in different forms by different authors (cf. [Lu, Chapter 4] and the references therein). For a k-regular finite graph X, we write $\lambda_1^*(X) = \frac{\lambda_1(X)}{k}$, where $\lambda_1(X)$ is the second to largest eigenvalue of X.

PROPOSITION.

$$\frac{2(1 - \lambda_1^*(X))}{2(1 - \lambda_1^*(X)) + 1} \le C(X) \le 2k\sqrt{1 - \lambda_1^*(X)}.$$

It follows that for a fixed k, a family $\{X_i\}$ of k-regular graphs is a family of expanders if and only if $\lambda_1(X)$ is uniformly bounded away from k. On the other hand if k is unbounded, the eigenvalue $\lambda_1(X)$ does not necessarily gives the complete answer: if $\lambda_1(X) \le (1 - \varepsilon)k$ for some $\varepsilon > 0$ then the graphs are indeed expanders (but $\lambda_1(X) < k - \varepsilon$ is not enough). In practice, the eigenvalues by themselves are not enough and other methods have to be invented (very few and very restrictive methods are known so far). See more in Chapter 3.

Finally we mention that the eigenvalues give a pretty good control of the rate of convergence of the random walk on Cayley graphs. This discussion is postponed to Chapter 4.

§2. CAYLEY GRAPHS OF BOUNDED DEGREE

(2.1)   Let $\{G_i\}_{i \in I}$ be a family of finite groups and for each $i \in I$, let Σ_i be a symmetric set of generators for G_i. **In this section we assume that $|\Sigma_i| = k$ for every i.**

It follows immediately from (1.11), that $\{X_i = X(G_i; \Sigma_i)\}_{i \in I}$ is a family of expanders if and only if $\lambda_1(X_i)$ are uniformly bounded away from k, i.e., there exists an $\varepsilon > 0$ such that for every i, $\lambda_1(X_i) < k - \varepsilon$. What is less clear is when and why this happens. In [LW, §3], it is shown that for various families of groups - e.g., G_i abelian or more generally solvable of bounded derived length, no choice of Σ_i would turn them into expanders. On the other hand various constructions of Cayley graphs which are expanders have been presented in the literature. Each case required a quite deep mathematical tool such as (1) Kazhdan property (T) from representation theory of semi-simple Lie groups, (2) Selberg's theorem estimating the eigenvalues of the Laplacian operator acting on congruence quotients of the hyperbolic upper half plane, (3) Ramanujan conjecture (as proved by Eichler and Deligne) estimating the Fourier coefficients of some cusp forms, or (4) Drinfeld's theorem which is the characteristic p analogue of Ramanujan conjecture.

What is very frustrating is that all these deep theories give some examples with very special sets of generators. A small change of the construction - which seems to be meaningless from the combinatorial point of view - leaves these tools helpless. Let's illustrate this by an example.

(2.2)   For a prime $p \geq 5$, let's define

$$\Sigma_p^0 = \left\{ \begin{pmatrix} 1 & 1 \\ 0 & 1 \end{pmatrix}, \begin{pmatrix} 0 & 1 \\ -1 & 0 \end{pmatrix} \right\}$$

$$\Sigma_p^1 = \left\{ \begin{pmatrix} 1 & 1 \\ 0 & 1 \end{pmatrix}, \begin{pmatrix} 1 & 0 \\ 1 & 1 \end{pmatrix} \right\}$$

$$\Sigma_p^2 = \left\{ \begin{pmatrix} 1 & 2 \\ 0 & 1 \end{pmatrix}, \begin{pmatrix} 1 & 0 \\ 2 & 1 \end{pmatrix} \right\}$$

$$\Sigma_p^3 = \left\{ \begin{pmatrix} 1 & 3 \\ 0 & 1 \end{pmatrix}, \begin{pmatrix} 1 & 0 \\ 3 & 1 \end{pmatrix} \right\}$$

and for $i = 0, 1, 2, 3$ let $X_p^i := X(SL_2(p); \Sigma_p^i)$.

PROPOSITION 2.2.1. *The family* $\{X_p^1\}_{p \geq 5}$ *is a family of expanders.*

PROOF: This follows from Theorem 4.4.2 in [Lu]. It is shown there that $\{X_p^0\}_{p \geq 5}$ form a family of expanders. The crucial point is that $\begin{pmatrix} 1 & 1 \\ 0 & 1 \end{pmatrix}$ and $\begin{pmatrix} 0 & 1 \\ -1 & 0 \end{pmatrix}$, as elements of the **infinite** modular group $\Gamma = SL_2(\mathbb{Z})$, generate Γ. Moreover, X_p^0 is a "discrete approximation" of the hyperbolic manifolds $\Gamma(p) \backslash \mathbb{H}$ where $\mathbb{H}$ is the upper-half plane and $\Gamma(p) = \mathrm{Ker}(SL_2(\mathbb{Z}) \to SL_2(p))$. Selberg's Theorem $\lambda_1(\Gamma(p) \backslash \mathbb{H}) \geq \frac{3}{16}$ is then used to prove that X_p^0 are expanders.

Now, $\begin{pmatrix} 1 & 1 \\ 0 & 1 \end{pmatrix}$ and $\begin{pmatrix} 1 & 0 \\ 1 & 1 \end{pmatrix}$ also generate Γ. Hence $\begin{pmatrix} 1 & 1 \\ 0 & 1 \end{pmatrix}$ and $\begin{pmatrix} 0 & 1 \\ -1 & 0 \end{pmatrix}$ can be presented as words in $\begin{pmatrix} 1 & 1 \\ 0 & 1 \end{pmatrix}$ and $\begin{pmatrix} 1 & 0 \\ 1 & 1 \end{pmatrix}$ of **bounded** length (independent of p). It is easy to see that this ensures that X_p^1 are also expanders.   Q.E.D.

On the other hand the subgroup of $\Gamma = SL_2(\mathbb{Z})$ generated by $\begin{pmatrix} 1 & 2 \\ 0 & 1 \end{pmatrix}$ and $\begin{pmatrix} 1 & 0 \\ 2 & 1 \end{pmatrix}$ is not Γ. Thus Σ_p^0 can not be expressed as words of bounded length using Σ_p^2. Still:

PROPOSITION 2.2.2. $\{X_p^2\}_{p \geq 5}$ *is a family of expanders.*

PROOF: Indeed, the subgroup Δ of Γ generated by $\begin{pmatrix} 1 & 2 \\ 0 & 1 \end{pmatrix}$ and $\begin{pmatrix} 1 & 0 \\ 2 & 1 \end{pmatrix}$ is of finite index in Γ. Selberg's theorem works for Δ and can be applied as

in the proof of Theorem 4.4.2 of [Lu] to give expanders with **some** generators of Δ. We can then deduce (as in Proposition 2.2.1) that by replacing
those generators with other generators of Δ, the quotients $SL_2(p)$ are also
expanders. Q.E.D.

On the other hand:

OPEN PROBLEM 2.2.3: Is $\{X_p^3\}_{p\geq 5}$ a family of expanders?

The difference is that $\begin{pmatrix} 1 & 3 \\ 0 & 1 \end{pmatrix}$ and $\begin{pmatrix} 1 & 0 \\ 3 & 1 \end{pmatrix}$ generate a subgroup of
infinite index in Γ. So, while they generate $SL_2(p)$ for every $p \geq 5$, the
standard generators (Σ_p^0) can not be expressed as words of bounded length
using Σ_p^3, nor can be any system of generators of any finite index subgroup
of Γ.

It is however difficult to believe that $\{X_p^3\}_{p\geq 5}$ are not expanders while
X_p^1 and X_p^2 are.

(2.3)   We do not even know the answer to the following questions (asked
in [Lu] and [LW]).

PROBLEM: (a) Are there systems of generators Σ_p for $SL_2(p)$ for which
$X(SL_2(p); \Sigma_p)$ are not expanders?

(b) Are there bounded systems of generators Σ_n for $SL_n(2)$ for which
$X(SL_n(2); \Sigma_n)$ are expanders?

(2.4)   Questions like that are of special interest for the symmetric
groups.

PROBLEM: Is there a bounded system of generators Σ_n for S_n for which
$X(S_n; \Sigma_n)$ are expanders?

If the answer is yes, then such generators can be used for generating
pseudo-random permutations.

(2.5)   It is known and easy to prove (see [Lu, p. 51] for three different proofs) that S_n are not expanders with respect to $\Sigma'_n = \{(1,2),(1,2,3,\ldots,n)\}$. It is not known whether groups can be expanders and non-expanders at the same time.

PROBLEM: Is there a family of finite groups G_n with two systems of generators Σ_n and Σ'_n such that: (a) $|\Sigma_n|,|\Sigma'_n| \leq k$ for some k, (b) $X(G_n,\Sigma_n)$ are expanders, and (c) $X(G_n,\Sigma'_n)$ are not expanders.

A negative answer to Problem 2.5 implies similar answer to Problems 2.4 and 2.3.

(2.6) It is also not known how k-random generators for $\{S_n\}_{n\geq5}$ or $\{SL_2(p)\}_{p\geq5}$ behave. Recently Liebeck and Shalev [LS] proved for the exceptional groups, the result below (which was proved before in [Di] for the alternating groups and in [KL] for the classical groups). So altogether:

THEOREM. *Let G be a finite simple group. Then the probability that two random elements generate G goes to 1, when $|G| \to \infty$.*

Essentially nothing is known about the expansion properties of the associated Cayley graphs (or even about their diameters - see [BHKLS]).

§3. CAYLEY GRAPHS OF UNBOUNDED DEGREE: EXPANSION

(3.1)   In the previous chapter we saw how little we know about expansion properties on Cayley graphs of bounded degree. For unbounded degree, the situation is not much better. Still, some progress has been made recently which deserves some attention.

(3.2)   First, the main problem concerning random generators was solved:

THEOREM (ALON-ROICHMAN [AR]). *For every $0 < \varepsilon < 1$ there exists a $c(\varepsilon) > 0$ such that the following holds: Let G be a group of order n, and let Σ be a random set of $c(\varepsilon)\log_2 n$ elements of G. Then the Cayley graph*

$X(G; \Sigma)$ is an ε-expander almost surely (i.e., the probability it is such an expander goes to 1 as n tends to infinity).

(3.3)   It should be pointed out that if one considers **all** finite groups - then Theorem 3.2 is best possible: if for abelian groups B_n one takes $o(\log |B_n|)$ generators, then the Cayley graphs are not expanders (see [AR]). If the groups B_n have bounded exponent, $o(\log |B_n|)$ elements do not even generate and the Cayley graphs are not connected.

(3.4)   For S_n, Theorem 3.2 says that $cn \log n$ generators suffice. It is no known whether S_n can be made into a family of expanders using $O(n)$ generators. (compare 2.4).

A related problem is the following:

Let $X(n, k, r)$ be the **random** graph obtained in the following way: The vertices are the ordered subsets of $I_n = \{1, \ldots, n\}$ consisting of r elements. We pick k random permutations $\sigma_1, \ldots, \sigma_k$ from S_n (the group of permutations of I_n) and we connect the ordered subset $(j_1, \ldots, j_r)$ with $(\sigma_i(j_1), \ldots, \sigma_i(j_r))$. (The graph $X(n, k, r)$ can be identified with the Schreier graph of the cosets of S_{n-r} in S_n; see §5). It is well known and easy to prove (cf. [Lu, Prop. 1.2.1]) that $X(n, k, 1)$ is expander with probability approaching 1. The other extremal case $X(n, k, n)$ coincides with a random Cayley graph on S_n with k generators. Whether these are expanders is open for k fixed (see 2.4 and 2.6) and even for $k = O(n)$ (see 3.4). On the other hand $X(n, cn \log n, n)$ are expanders by Theorem 3.2. It will be interesting to understand where the "cut off points" are.

Here is a partial result:

THEOREM [FJRST]. *For every fixed r there exists k such that for $n \to \infty$, most graphs $X(n, k, r)$ are expanders.*

(3.5)   We now turn to consider Cayley graphs of finite groups with respect to some special (unbounded) sets of generators. Here is a case with an amazingly precise answer:

THEOREM (BACHER-DE LA HARPE [BDH] BACHER [Ba]). *Let* $X(n) = X(S_n; \Sigma_n = \{(1,2),(2,3),\ldots,(n-1,n)\})$ *be the* $(n-1)$-*regular Cayley graph of* S_n *with respect to the Coxeter generators. Then: (a) The Kazhdan constants of* S_n *with respect to these generators are* $\hat{K}_{S_n}(\Sigma_n) = K_{S_n}(\Sigma_n) = \sqrt{\frac{24}{n^3-n}}$
        (b) $\lambda_1(X(n)) = n - 3 + 2\cos\frac{\pi}{n}$.

Recall that the Kazhdan constants of a group G with respect to a set of generators Σ are defined as follows:

$$K_G(\Sigma) = \inf\{K_G(\Sigma, \rho) \mid \rho \text{ a unitary representation without a fixed vector}\}$$
$$\hat{K}_G(\Sigma) = \inf\{K_G(\Sigma, \rho) \mid \rho \text{ an irreducible non trivial unitary representation}\}$$

where for a unitary representation ρ on an Hilbert space H_ρ one defines:

$$K_G(\Sigma, \rho) = \inf_{\substack{v \in H_\rho \\ ||v||=1}} \max_{\sigma \in \Sigma} ||\rho(\sigma)v - v||.$$

(3.6)   As it is well known the expansion constant of the Cayley graph is closely related to the Kazhdan's constants of the group (see [BLH, Proposition 6]). In the case under consideration, $X(S_n; \Sigma_n)$, it follows that these graphs are not expanders. This last result is also a corollary of (3.7b) below.

(3.7)   On the other hand the Cayley graphs of the symmetric groups with respect to the conjugacy class of long cycles ($\geq \Omega(n)$) are expanders! More precisely, let $C_{f(n)}$ be the conjugacy class in S_n of all cycles of length $f(n)$. Let $\langle C_{f(n)} \rangle$ be the group generated by it. Clearly, $\langle C_{f(n)} \rangle$ is equal to S_n if $f(n)$ is even and to A_n if $f(n)$ is odd.

PROPOSITION (ROICHMAN [R4]). *(a) If* $f(n) = \Omega(n)$, *i.e., there exists* $\delta > 0$ *such that* $f(n) \geq \delta n$, *then* $X(\langle C_{f(n)}\rangle; C_{f(n)})$ *is a family of expanders.*

(b) If $f(n) = o(\sqrt{n})$ *then* $X(\langle C_{f(n)}\rangle; C_{f(n)})$ *are not expanders.*

(3.8)   *It is not known what is the situation if* $f(n) = \Omega(\sqrt{n})$ *but also* $f(n) = o(n)$. *In fact,*

PROPOSITION. *Let* $\lambda_1^*(n) = \lambda_1^*(X\langle C_{f(n)}\rangle; C_{f(n)})$ *be the normalized second to the largest eigenvalue of* $X = X(\langle C_{f(n)}\rangle; C_{f(n)})$, *i.e., the second to the largest eigenvalue of* X *divided by the degree of* X *(see 1.11). Then:*

(a) *If* $f(n) = \Omega(n)$ *then there exists* $q < 1$ *such that* $\lambda_1^*(n) < q$ *for every* n.

(b) *If* $f(n) = o(n)$ *then* $\lim_{n \to \infty} \lambda_1^*(n) = 1$.

(3.9)   The Proposition is proved using character estimates on the symmetric group (see §4.6 below for a detailed explanation). Anyway, Proposition 3.8(a) together with Proposition (1.11) imply Proposition 3.7(a). On the other hand Proposition 3.8(b) does not enable us to decide whether $X(\langle C_{f(n)}\rangle; C_{f(n)})$ are expanders for $f(n) = o(n)$. They are not, as said above, for $f(n) = o(\sqrt{n})$ by a direct combinatorial argument. Roichman conjectures that they are expanders for $f(n) = \Omega(\sqrt{n})$. he proved a partial result in this direction:

THEOREM ([R4]). *For every* $0 < \delta < 1$ *and every function* $f(n) > \sqrt{2n}$, *there exists a constant* c *such that* $|\partial(A)| \geq c|A|$ *for every* $A \subseteq X(\langle C_{f(n)}\rangle; C_{f(n)})$ *whose size is at most* $(1 - \delta)\left(n - \frac{n}{f(n)}\right)!$. *I.e., these graphs expand small sets.*

While the result is weaker than what is desired it has an interesting proof: Two methods are usually used for proving expansion - upper bound on eigenvalues (1.11) or small diameter (3.13). None of them works in this case. Roichman presents the regular representation of $G = \langle C(n, r)\rangle$ as a sum of two representations. In one, the eigenvalues method works. The complement has a combinatorial interpretation as a permutational representation in which the diameter method applies. Together the theorem is deduced.

See also Schechtman [Sc] and the references therein for some related results.

(3.10)   It is sometimes useful to think of S_n as the general linear group, $GL_n(\mathbb{F}_1)$, over "the field" of order 1. In our context, however, it seems that the simple groups of Lie type behave quite differently from the family S_n (or A_n). For S_n and A_n we saw that they are expanders with respect to some conjugacy classes and they are not with respect to others. On the other hand for simple groups of Lie type we always get expanders (provided the field of definition is large enough and probably also without this restriction).

THEOREM. *There exists a constant M such that: let $\{G_n(\mathbb{F}_{q_n}\}_{n \in \mathbb{N}}$ be a family of finite simple groups of Lie type, where G_n is defined in its natural definition over the field of order q_n. Let C_n be an arbitrary non-trivial conjugacy class of $G_n(\mathbb{F}_{q_n})$ and $X_n = X(G_n(\mathbb{F}_{q_n}); C_n)$. Then if for all n, $q_n \geq M$, then $\{X_n\}_{n \in \mathbb{N}}$ forms a family of expanders.*

The Theorem follows from results of Gluck ([Gl1], [Gl2]): Indeed as explained below in §4.6, the normalized eigenvalues of the Cayley graph of a group G with respect to a conjugacy class C are the normalized characters $\chi(C)/\chi(e)$, when e is the identity element and χ runs over the irreducible characters of G. Theorem 3.11 below implies that for q sufficiently large $\chi(C)/\chi(e)$ is uniformly bounded away from 1. Hence (1.11) shows that these are all expanders which proves Theorem 3.10.

(3.11) THEOREM (GLUCK). *There exists a constant M such that for every finite simple group of Lie type $G(q)$, every non-trivial element $g \in G(q)$ and* *every non-trivial irreducible character* χ *of* $G(q)$, $|\chi(g)/\chi(e)| < \frac{M}{q^{1/2}}$.

It worth mentioning (compare 4.12 below) that if we bound the Lie rank of the finite simple groups and let q go to infinity then $\chi(C)/\chi(e)$ (and hence the eigenvalues) go to zero. On the other hand if we take a large but fixed q and look at $PSL_n(q)$ with $n \to \infty$, then $\chi(C)/\chi(e)$ is bounded away from 1, but does not go to zero.

(3.12)   We are left with the open question whether for small q, $G_n(q)$ are always expanders with respect to conjugacy classes. It is very likely that this is the case (provided $q \geq 2$, of course).

(3.13)   Being expanders is related to diameters: Expanders have small diameters - i.e., diameter$(X_n) = O(\log |X_n|)$ for a family $\{X_n\}$ of expander graphs. But, usually, small diameter does not ensure expansions. There is one result in this direction:

THEOREM ([Al], [Bab]). *If* $X = X(G, \Sigma)$ *is a Cayley graph, then* $C(X) \geq$ $(2\text{diameter}(X))^{-1}$.

This implies that Cayley graphs with bounded diameters are expanders. (This gives a proof of Proposition 3.7(a) without eigenvalues). The diameters of simple groups with respect to conjugacy classes were studied extensively (see Arad-Herzog [AH] and the references therein). It is known (combine for example Proposition 4.5 of [AFM] with Gluck's theorem (3.11) above) that for simple groups of Lie type of bounded rank the diameter is bounded for all non trivial conjugacy classes.

§4. RANDOM WALKS ON CAYLEY GRAPHS

(4.1)   Consider the random walk on a finite graph X in which every step consists of moving with probability $\frac{1}{k}$ along one of the k edges coming out of the vertex. The random walk defines a Markov chain whose possible states are the vertices of X and a Markov operator $M : L^2(X) \to L^2(X)$ given by: $(Mf)(x) = \frac{1}{k} \sum_{y \in X} \delta_{xy} f(y)$, when δ is the adjacency matrix of X. So M is nothing but $\frac{1}{k} \delta$ where δ is the adjacency matrix of X. If P is an initial probability distribution on X, then in the n-th step of the random walk the distribution will be $M^n(P)$.

The vector $U = (\frac{1}{n}, \ldots, \frac{1}{n})$ when $n = |X|$, represents the uniform distribution. If X is connected and not bi-partite then for every probability vector P, $\lim M^i(P) = U = \tilde{U}$. If $X = X_1 \cup X_2$ is connected and bi-partite then the random walk does not converge to the uniform distribution since it goes alternatingly between X_1 and X_2. Still $\lim M^{2i}(P) = \varepsilon U_1 + (1 - \varepsilon)U_2 = \tilde{U}$ when $\varepsilon = \sum_{x \in X_1} P(x)$ and U_i is the distribution which is zero on X_{3-i} and uniform on X_i.

(4.2)   Let now $X = X(G; \Sigma)$ be the Cayley graph of a finite group with respect to a symmetric set of generators Σ with $|\Sigma| = k$. X is always connected. It is bi-partite if and only if G has a normal subgroup H of index 2 and $\Sigma \cap H = \emptyset$, in which case the two parts of X are $X_1 = H$ and $X_2 = G \setminus H$. Let

$$Q(g) = \begin{cases} \frac{1}{k} & g \in \Sigma \\ 0 & g \notin \Sigma. \end{cases}$$

Then Q is a probability distribution on G and in the notations of (4.1),

$M^n \begin{pmatrix} 1 \\ 0 \\ \vdots \\ 0 \end{pmatrix} = Q^{*n}$ when $\begin{pmatrix} 1 \\ 0 \\ \vdots \\ 0 \end{pmatrix}$ denotes the probability vector giving e (the identity of G) measure 1 and 0 elsewhere, and Q^{*n} denotes Q convolved with itself n times. (Recall that for two functions φ and ψ on G, the convolution is defined as $\varphi * \psi(x) = \sum_{y \in G} \varphi(xy^{-1})\psi(y)$.)

(4.3)   Our main problem is to understand how fast this random walk on $X = X(G; \Sigma)$ converges to the uniform distribution? In technical terms: given $\varepsilon > 0$, how large should t be so that $||Q^{*t} - U|| < \varepsilon$?

A word of explanation is needed about the norm $||\ ||$ we are using: For various good reasons mathematicians usually prefer L^2-norm which is usually more tractable. But, as advocated convincingly by Diaconis [D, Chapter 3] for random walks the **variation distance** between two probability distributions is more natural and useful.

DEFINITION: Let P and Q be two probability distributions on a finite set X. Define the **variation distance** between P and Q as:

$$||P - Q|| = \max_{A \subseteq X} |P(A) - Q(A)|.$$

It is easy to prove that

$$||P - Q|| = \frac{1}{2} \sum_{x \in X} |P(x) - Q(x)|.$$

Thus, $||P - Q|| = \frac{1}{2}||P - Q||_1$ when $||P - Q||_p$ denotes the L^p-norm. So, essentially $||P - Q||$ is the L^1-norm.

(4.4)   The variation distance is bounded by the L^2-norm: $||P - Q|| \leq \frac{1}{2}|X|^{1/2}||P - Q||_2$. It is therefore still useful to consider also the L^2-norm. For it, we have the following easy proposition:

PROPOSITION. *Let $X(G; \Sigma)$ be as before when G is a group of order g. Let Q and M be as in (4.2) and $\lambda_0 = 1 > \lambda_1 \geq \lambda_2 \geq \ldots \geq \lambda_{g-1} \geq -1$ the eigenvalues of M. Then:*

$$||Q^{*t} - U||_2^2 = Q^{*2t}(e) - \frac{1}{g} = \frac{1}{g}(\text{tr}(M^{2t}) - 1) = \frac{1}{g}\sum_{1}^{g-1} \lambda_i^{2t}$$

PROOF:

$$\|Q^{*t} - U\|_2^2 = \sum_{x \in X} \left(Q^{*t}(x) - \frac{1}{g} \right)^2$$

$$= \sum_{x \in X} Q^{*t}(x)^2 - 2\frac{1}{g} \sum_{x \in X} Q^{*t}(x) + g \cdot \frac{1}{g^2}$$

$$= \sum_{x \in X} Q^{*t}(x) Q^{*t}(x^{-1}) - \frac{1}{g} =$$

$$= Q^{*2t}(e) - \frac{1}{g}.$$

On the other hand $Q^{*2t}(e)$ is equal to $\frac{1}{k^{2t}}$ times the number of closed paths on X from e to e of length $2t$, where e is, as usual, the identity element of G. This number is independent of the base point and equals to $\frac{1}{g}\mathrm{tr}(M^{2t})$. Thus $Q^{*2t}(e) - \frac{1}{g} = \frac{1}{g} \sum_{1}^{g-1} \lambda_i^{2t}$.   Q.E.D.

Of course, if we want a bound on $\| \ \|_2$ we can replace the sum $\sum_{1}^{g-1} \lambda_i^{2t}$ by $\frac{|G|}{k^{2t}}\lambda_1(X)^{2t}$ (when $\lambda_1(X)$ is the second to largest eigenvalue of X), but this might be a too crude estimate. So, unlike the expansion property which depends almost entirely on $\lambda_1(X)$, for the mixing problem the other eigenvalues are also important. Usually, it is easier to estimate the sum of the eigenvalues than estimating their maximum.

COROLLARY. $\|Q^{*t} - U\|^2 \leq \frac{1}{4} \sum_{1}^{g-1} \lambda_i^{2t} \leq \frac{|G|}{4} \left(\frac{\lambda_1(X)}{k} \right)^{2t}$.

(4.5)   When $X = X(G; \Sigma)$ is a Cayley graph, the operator M has an additional interpretation: Identify $L^2(X)$ with the group algebra $\mathbb{C}[G]$ and then M acts on $L^2(X)$ exactly as $\frac{1}{k} \sum_{s \in S} s \in \mathbb{C}[G]$ acts on $\mathbb{C}[G]$ by multiplication from the right. As it is well known (cf. [D]) $\mathbb{C}[G]$ is decomposed as $\oplus_{\rho} d_\rho V_\rho$ where (V_ρ, ρ) runs over the equivalence classes of irreducible representations of G and $d_\rho = \dim V_\rho$. Denote $\chi_\rho(g) = \mathrm{tr}\rho(g)$, Proposition (4.4) can be written as:

$$||Q^{*t} - U||_2^2 = \frac{1}{g}\sum_{\rho}{}^* d_\rho \chi_\rho(M^{2t})$$

when $\sum^*$ means that the sum runs only over the non-trivial irreducible representations ρ.

(4.6)   Representation theory of finite groups gives information on the characters χ_ρ, but usually it is impossible to estimate χ_ρ on **powers of** M. The seminal paper of Diaconis and Shahshahani [DS] illustrates how one can do it (and thus answers the "rate of mixing problem") in some cases. Here is the recipe: Assume Σ is a union of conjugacy classes (or even more generally Q is a probability measure which is constant on conjugacy classes). Then $M = \frac{1}{k}\sum_{s\in\Sigma} s$ commute with left multiplication on $\mathbb{C}[G]$ and it is therefore given by a scalar matrix on each V_ρ (Schur's lemma).

This scalar matrix is equal to $\frac{1}{k\cdot d_\rho}\sum_{s\in\Sigma}\chi_\rho(s)$. Assume further (for simplicity) that Σ is just one conjugacy class C, then this is equal to $\frac{|C|}{kd_\rho}\chi_\rho(C) = \frac{\chi_\rho(C)}{d_\rho}$. This last term $\tau_\rho(C) = \frac{\chi_\rho(C)}{d_\rho}$ is called the **normalized character**.

We thus have:

PROPOSITION (UPPER BOUND LEMMA (CF. [D, CHAP. 3B])). *Let C be a conjugacy class in a finite group G and Q the probability distribution whose support is uniformly on Σ. Then:*

$$||Q^{*t} - U||^2 \leq \frac{1}{4}\sum_{\rho}{}^* d_\rho^2 \tau_\rho(C)^{2t}$$

when the sum is over the non-trivial irreducible representations ρ of G.

PROOF: Follows from the first inequality of Corollary 4.4 and the discussion in (4.5) and (4.6).   Q.E.D.

Diaconis and Shahshahani [DS] used a formula of Frobenius to estimate $\tau_\rho(C)$ when C is the conjugacy class of the transpositions in the symmetric group S_n. They have deduced:

(4.7) THEOREM. *Let* Q *be the following distribution on* S_n: $Q(e) = \frac{1}{n}$, $Q(\tau) = \frac{2}{n^2}$ *if* τ *is a transposition and* $Q(\pi) = 0$ *otherwise. Let* $t = \frac{1}{2}n\log n + cn$. *Then*

(a) *for* $c > 0$, $\|Q^{*t} - U\| \leq ae^{-2c}$ *for a universal constant* a.

(b) *For* $c < 0$, *and* n *going to infinity:*

$$\|Q^{*t} - U\| \geq \left(\frac{1}{e} - e^{-e^{-2c}}\right) + o(1).$$

This Theorem has the pleasant interpretation: $\frac{1}{2}n\log n + O(n)$ random transpositions mix a deck of n cards pretty well.

The theorem also illustrates beautifully a phenomena of "cut off" which occurs frequently in these problems: the variation distance, as a function of t, is essentially 1 for a time and then rapidly becomes tiny and tends to zero exponentially fast past the cut off.

(4.8)   Recently, the problem was solved also for more general conjugacy classes of the symmetric groups. To save the bother of dealing with the bipartite case, we quote the theorem for the alternating groups A_n. Supp(C) means the number of non-fixed elements of a permutation in C.

THEOREM (ROICHMAN [R3]). *For a conjugacy class* C *in the alternating group* A_n, *let*

$$Q_C(g) = \begin{cases} \frac{1}{|C|} & g \in C \\ 0 & g \notin C. \end{cases}$$

Then for every $\delta > 0$ *and every conjugacy class* C *with support less then* $(1 - \delta)n$ *and for* n *sufficiently large, the rate of convergence of* Q_C *in* A_n *is* $\Theta\left(\frac{n\log n}{\operatorname{supp}(C)}\right)$, *i.e. given* $\varepsilon > 0$ *there exists constants* $a(\delta,\varepsilon)$ *and* $b(\delta,\varepsilon)$ *such that for* $t > a(\delta,\varepsilon)\frac{n\log n}{\operatorname{supp}(C)}$, $\|Q_C^{*t} - U\| < \varepsilon$ *and for* $t < b(\delta,\varepsilon)\frac{n\log n}{\operatorname{supp}(C)}$, $\|Q_C^{*t} - U\| > \varepsilon$.

So, up to evaluating the constants a and b, the theorem solves the problem for an arbitrary conjugacy class, whose support is less then $(1 - \delta)n$.

(4.9) An essential ingredient of the proof of Theorem 4.8 is the following deep estimate, also due to Roichman, on normalized characters.

THEOREM ([R2]). *There exist constants $c > 0$ and $0 < q < 1$, such that for every conjugacy class C in the symmetric group S_n and every irreducible representation ρ^λ;*

$$|\tau_\rho(C)| \leq \left(\max\left\{ \frac{\lambda_1}{n}, \frac{\lambda_1'}{n}, q \right\} \right)^{c \cdot \text{supp}(C)}$$

where λ is the partition of n corresponding to ρ^λ, λ_1 the first term of that partition (=the width of the associated Young diagram) and λ_1' is the number of terms in λ (=the height of the Young diagram).

(4.10)   Theorem 4.8 does not cover the case of conjugacy classes whose support is of size close to n. For some of these Lulov [Lul] have used the character estimates from Fomin-Lulov [FL] to get:

THEOREM. *For any sequence of conjugacy classes of A_n with a bounded number of fixed points the asymptotic mixing time is 2 or 3 steps.*

For a sequence of conjugacy classes C_n in A_n, we say that the asymptotic mixing time is ℓ if $||Q_{C_n}^{*\ell} - U|| \to 0$ as $n \to 0$.

For $n \equiv 0 \pmod 4$ and C_n the conjugacy class of fixed point free involutions, Lulov showed that 3 steps are needed. On the other hand for $r \geq 3$ and $n \equiv 0 \pmod r$, if C_n is the conjugacy class of $\frac{n}{r}$ r-cycles, 2 steps suffice.

(4.11)   In [Lu, Chapter 8.2] it was shown how the character table of $SL_2(q)$ gives full information on the eigenvalues of $X(SL_2(q); \Sigma)$ where Σ is any conjugacy class of $SL_2(q)$ (q is a prime power). One can easily calculate to deduce:

PROPOSITION. *For every $\varepsilon > 0$ there exists a constant $\ell = \ell(\varepsilon)$ such that for every q and every non-central conjugacy class C of $SL_2(q)$, $||Q_C^{*\ell} - U|| < \varepsilon$*

where

$$Q_C(g) = \begin{cases} \frac{1}{k} & g \in C \cup C^{-1} \\ 0 & g \notin C \cup C^{-1} \end{cases}$$

and $k = |C \cup C^{-1}|$.

(4.12)   Gluck [Gl1, Gl2] estimated the normalized characters for more general finite simple groups of Lie type. His Theorem 3.11 above, implies:

THEOREM. *For fixed r and ε, there exists a constant $t = t(r, \varepsilon)$ such that for every finite simple group of Lie type of rank $\leq r$ and every conjugacy class C in G, $\|Q_C^{*t} - U\| < \varepsilon$.*

PROOF: By Corollary 4.4, $\|Q^{*t} - U\|^2 \leq \frac{1}{4} \sum_1^{g-1} \lambda_i^{2t}$. By Theorem 3.11 for some fixed M, $|\lambda_i| < \frac{M}{q^{1/2}}$. Thus $\|Q^{*t} - U\|^2 \leq \frac{|G|}{4} \left(\frac{M}{q^{1/2}} \right)^{2t}$. Now, for a group $G = G(q)$ of bounded rank $|G| < q^c$ and hence one can find t independent of G, such that $\|Q^{*t} - U\| < \varepsilon$.   Q.E.D.

(4.13)   The dependence of t on r is necessary: If r is unbounded, say $G = SL_n(q)$ and $n \to \infty$, it is easy to see that the conjugacy class of the transvections (i.e., the non-trivial linear transformations which fix some hyperplane pointwise and have all eigenvalues 1) in $SL_n(q)$ needs at least $\Omega(n)$ steps to generate G, so there is no such fixed t. (See [AH] for a comprehensive study of the diameter of Cayley graphs with respect to conjugacy classes.) In fact, Hildebrand [Hi] determined precisely the mixing time of the random walk on $SL_n(\mathbb{F}_q)$ with respect to transvections.

THEOREM. *Let $X_n^q = X(SL_n(\mathbb{F}_q); C_n^q)$ be the Cayley graph of $SL_n(\mathbb{F}_q)$ with respect to transvections. Given $\varepsilon > 0$, there is a constant c such that after $n + c$ steps the random walk on X_n^q is within a distance ε from uniform and after $n - c$ steps the walk is a distance at least $1 - \varepsilon$ from uniform.*

Also here the main part of the proof is an estimate of the normalized characters on the conjugacy class C_n^q.

An interesting problem is to find the analogue to Theorem 4.8 for groups of Lie type. It might be that there is an interesting difference between S_n and groups of Lie type: For $G_n = S_n$ and Σ_n the conjugacy class of all transpositions the diameter of $X(G_n; \Sigma_n)$ is $O(n)$ while the mixing time is $\frac{1}{2}n \log n + O_\varepsilon(n)$. On the other hand for groups of Lie type it might be that the following question has an affirmative answer:

PROBLEM: Given $\varepsilon > 0$. Is there a constant $c(\varepsilon)$ such that for every finite simple group of Lie type G and every conjugacy class C in G, $\|Q_C^{*t} - U\| < \varepsilon$ for $t = c(\varepsilon)$ diameter $X(G; C)$?

It will be also of interest to understand if there is a difference between groups of bounded and unbounded rank (compare [LW] and [GKS]).

To keep the right perspective we must add that the question of rate of mixing is a central one and many papers have been devoted to it applying diverse interesting techniques. Our main interest in this survey is to get some insight into the combinatorics of the Cayley graphs of finite (simple) groups, so our selection was far from being representative. For an excellent treatment of random walks on finite groups and for many more references, the reader is referred to [D], [Al] and [FOW].

Recently, U. Porod ([P1], [P2]) proved some interesting results of the kind described in this chapter for compact simple Lie groups. An interesting new phenomenon occurs there: while in the finite case there is only one interesting distribution on a conjugacy class (namely, the uniform one), in her context there are more. The mixing time turns out to be dependent on the starting distribution on the conjugacy class.

§5. SCHREIER GRAPHS AND ISOSPECTRAL GRAPHS

(5.1)   Two finite graphs of order n, X_1 and X_2 are called **isospec-tral** (or: **cospectral**) if their spectrum (including multiplicities of eigen-

values) are the same. Several methods for constructing isospectral graphs
are known (cf. [CDS] and [CDGT]). In this section we describe an additional one. This method is a small variant of the method used by Sunada
([Su]) to present isospectral Riemannian manifolds, i.e., pairs of Riemannian
manifolds with the same spectrum of their Beltrami-Laplace (=Laplacian)
operators. Sunada's method has been used recently ([GWW]) to answer the
long outstanding problem of M. Kac: Can you hear the shape of a drum?

   One certainly cannot hear the shape of a graph!

(5.2)   For constructing isospectral graphs we will use **Schreier graphs**.
These are quotients of Cayley graphs. We take the opportunity of putting
on record some easy results related to Schreier graphs which seems not to
be as well known as they should.

DEFINITION: Let G be a group, H a subgroup of G and Σ a sequence
$\sigma_1, \sigma_1^{-1}, \ldots, \sigma_s, \sigma_s^{-1}, \tau_1, \ldots, \tau_t$ of elements of G such that the τ_i $(1 \leq i \leq t)$
are of order ≤ 2 and σ_j $(1 \leq j \leq s)$ are of order ≥ 3. Assume Σ generates
G. Let $X(G/H; \Sigma)$ be the graph whose vertices are the left cosets of H in
G. For $gH \in G/H$ and $\rho \in \Sigma$ we put an edge from gH to $\rho g H$. (If $\rho = \sigma_i^{\pm 1}$
for some $1 \leq i \leq s$ then this edge is identified with the edge corresponding
to ρ^{-1} from $\rho h H$ to gH).

   The graph $X(G/H; \Sigma)$ is called the **Schreier graph** of G/H with respect to Σ.

(5.3) REMARK: (a) If H is normal in G, $X(G/H; \Sigma)$ is the Cayley graph of
G/H with respect to the image of Σ in G/H. Formally speaking we defined
Cayley graphs in (0.1) with respect to a **subset** Σ and not like here with
respect to a sequence Σ which might contain repetition. But, clearly the
definition of Cayley graphs can be extended to sequences. This is not so
interesting for Cayely graphs. For Schreier graphs, it is worthwhile allowing
this in order to make Theorem 5.4 below correct.

(b) $X(G/H; \Sigma)$ might have loops and double edges (even if Σ contains no repetitions). Indeed $\rho \in \Sigma$ define a loop in gH iff $g^{-1}\rho g \in H$. Two elements ρ_1, ρ_2 define double edge from gH to $\rho_1 gH = \rho_2 gH$ iff $g^{-1}\rho_2^{-1}\rho_1 g \in H$.

(c) The Schreier graph $X(G/H; \Sigma)$ is the quotient graph of the Cayley graph $X(G; \Sigma)$ acted upon from the right by H. But a caution is needed: The action of H on $X(G; \Sigma)$ is **without inversion** (in the sense of [Se, p. 25]) if and only if the following condition is satisfied:

(∗) For every $1 \leq i \leq t$ and every $g \in G$, $g^{-1}\tau_i g \notin H$. Namely, H contains no conjugate of the elements of order 2 in Σ.

(d) If condition (∗) is satisfied then $X(G/H; \Sigma)$ is a k-regular graph for $k = 2s + t$. On the other hand if (∗) is not satisfied then $X(G/H; \Sigma)$ is not k-regular: if $g^{-1}\tau_i g \in H$ then $\tau_i gH = gH$ and so τ_i defines a loop at gH, hence adds two to its degree. (See Example 5.12 below). One may still think of $X(G/H; \Sigma)$ as k-regular with one "folded" edge attached to gH. This is the "correct" geometrical picture, but it is not a graph any more.

(e) In our definition we assume that Σ generates G. It is actually possible that $X(G/H; \Sigma)$ is connected even if Σ does not generate G. But in such a case $X(G/H; \Sigma)$ is isomorphic to $X(K/H \cap K; \Sigma)$ where K is the subgroup generated by Σ. So, we may as well assume Σ generates G.

(5.4)   Regular graphs are usually Schreier graphs:

THEOREM. *Let X be a finite connected k-regular graph, $2 \leq k \in \mathbb{N}$.*

(a) *(Gross [Gs]) If k is even, then X is isomorphic to a Schreier graph.*

(b) *If k is odd then X is isomorphic to a Schreier graph if X has 1-factor, i.e., if there is a subset L of the set of edges such that every vertex lie on a unique edge of L.*

PROOF: (a) Assume the n vertices of X are $I = \{1, 2, \ldots, n\}$. By a theorem of Peterson (cf. [Gs]) every regular graph (connected or not) of even degree has a 2-factor, i.e., a subset of edges which forms on the set of vertices of X, a 2-regular subgraph. It follows that the edges of X can be partitioned into 2-factors $F_1, \ldots, F_r$, $r = k/2$. Assign an arbitrary orientation to each cycle of the 2-factors. By doing this every F_i defines a unique permutation π_i of I, whose cycles are the cycles of F_i. Let G be the subgroup of S_n generated by $\pi_1, \ldots, \pi_r$ and H the subgroup of G fixing the vertex 1. Then X is isomorphic to $X(G/H; \{\pi_1, \pi_1^{-1}, \ldots, \pi_r, \pi_r^{-1}\})$.

(b) Say $k = 2r + 1$ and assume X has a 1-factor F_0. Delete this 1-factor and on the resulting even regular graph apply the proof of (a) to get permutations $\pi_1, \ldots, \pi_r$. The 1-factor F_0 defines an involution τ of $X = I = \{1, \ldots, n\}$. Then X is isomorphic to $X(G/H; \{\pi, \pi_1^{-1}, \ldots, \pi_r, \pi_r^{-1}, \tau\})$ when G is the subgroup of S_n generated by $\pi_1, \ldots, \pi_r, \tau$ and H is the stabilizer of 1.   Q.E.D.

(5.5) REMARKS: (a) If X is a Schreier graph (as in (5.2)) of odd degree $k = 2s + t$ and if it satisfies condition $(*)$ of 5.3(c) (actually, if it satisfies it for **one** i, $1 \leq i \leq t$) then X has 1-factor.

(b) The well known Tutte's 1-factor theorem [Bo, p. 59] gives a necessary and sufficient condition for a graph to have a 1-factor.

(5.6) DEFINITION: Let G be a finite group and H_1, H_2 two subgroups of G. The triple $(G; H_1, H_2)$ will be called a **Sunada pair** if for every conjugacy class C of G, $|C \cap H_1| = |C \cap H_2|$.

(5.7) PROPOSITION. *Let G be a finite group and H_1, H_2 two subgroups. Then:*

(a) *H_1 is conjugate to H_2 if and only if the two **permutational** representations of G on G/H_1 and on G/H_2 are equivalent.*

(b) *$(G; H_1, H_2)$ is a Sunada pair if and only if the two **linear** representations of G on $L^2(G/H_1)$ and $L^2(G/H_2)$ are equivalent.*

PROOF: (a) follows from the fact that in transitive actions the stabilizers of different points are conjugate. For (b) recall that two linear representations ρ_1 and ρ_2 are equivalent if and only if their characters χ_{ρ_1} and χ_{ρ_2} are the same. For the linear representation ρ_i of G on $L^2(G/H_i)$ obtained from the permutational representation of G on G/H_i, we have:

$$\chi_{\rho_i}(g) = \#\text{ fixed points of } g \text{ acting on } G/H_i =$$
$$= \frac{1}{|H_i|}|\{t \in G \mid gtH_i = tH_i\}| =$$
$$= \frac{1}{|H_i|}|\{t \in G \mid t^{-1}gt \in H_i\}| =$$
$$= \frac{|Z_G(g)|}{|H_i|}|C_g \cap H_i|$$

where $Z_G(g)$ is the centralizer of g in G and C_g is the conjugacy class of g. This shows that $\chi_{\rho_1} = \chi_{\rho_2}$ if and only if $(G; H_1, H_2)$ is a Sunada pair. Q.E.D.

(5.8)   We named $(G; H_1, H_2)$ "Sunada pair" following the geometers, but one may argue whether this name is justified. This notion has been used by group theorists and number theorists for a long time (cf. [Ga]). For example, if F_1 and F_2 are two finite field extensions of the field of rational numbers $\mathbb{Q}$, then the Dedekind zeta function $\zeta_{F_1}(s)$ of F_1 is equal to $\zeta_{F_2}(s)$ if and only if $(\mathrm{Gal}(F/\mathbb{Q}); \mathrm{Gal}(F/F_1), \mathrm{Gal}(F/F_2))$ is a Sunada pair when F is a Galois extension of $\mathbb{Q}$ containing both F_1 and F_2 (cf. [CF, p. 363]). Actually, this fact was the starting point of Sunada's method. He (and others) have developed the notion of zeta function of a Riemannian manifold M encoding the spectrum of (the Laplacian of) M. He then used "Sunada pairs" to show examples of Riemannian manifolds with the same zeta functions and hence isospectral. Graphs also have zeta functions (cf. [Lu, §4.5]) and his method can be adapted also for them. Sunada's proof is pretty easy - but with graphs it is even easier to see it without zeta functions:

(5.9) THEOREM. *Let $(G; H_1, H_2)$ be a Sunada pair and Σ a set of generators of G. Then the Schreier graphs $X_1 = X(G/H_1; \Sigma)$ and $X_2 = X(G/H_2; \Sigma)$ are isospectral.*

PROOF: Let $W_\ell(\Sigma)$ be the set of (non reduced) words of length ℓ in Σ. Let δ_i be the adjacency matrix of X_i, $i = 1, 2$. X_1 and X_2 are isospectral if and only if for every $\ell \in \mathbb{N}$, $\mathrm{tr}(\delta_1^\ell) = \mathrm{tr}(\delta_2^\ell)$. Now,

$$\mathrm{tr}(\delta_i^\ell) = \sum_{x \in X_i} \left(\begin{array}{c} \text{\# closed paths on } X_i \text{ of length } \ell \\ \text{starting at } x \end{array} \right)$$

$$= \# \{(gH_i, w) \mid gH_i \in G/H_i, \ w \in W_\ell(\Sigma) \text{ and } wgH = gH_i\}$$

$$= \frac{1}{|H_i|} \# \{(g, w) \mid g \in G, \ w \in W_\ell(\Sigma) \text{ and } g^{-1}wg \in H_i\}$$

$$= \frac{1}{|H_i|} \sum_{w \in W_\ell(\Omega)} \# \{g \in G \mid g^{-1}wg \in H_i\}$$

$$= \frac{1}{|H_i|} \sum_{w \in W_\ell(\Omega)} (|Z_G(w)| \cdot |C_w \cap H_i|)$$

This shows that if $(G; H_1, H_2)$ is a Sunada pair then $\mathrm{tr}(\delta_1^\ell) = \mathrm{tr}(\delta_2^\ell)$. Q.E.D.

(5.10)   There are quite a lot of examples of Sunada pairs. here is one which is easy to demonstrate: Let p be a prime and H_1 and H_2 be two finite groups of order p^r and of exponent p (i.e., $x^p = 1$ for every $x \in H_i$, $i = 1, 2$). Each H_i is embedded in $G = S_{p^r}$ by its action on itself by right translations. $(G; H_1, H_2)$ is a Sunada pair. Indeed, every $1 \neq x \in H_i$ as an element of G is a product of p^{r-1} p-cycles, so all non-trivial elements of H_1 are conjugate in G to all non-trivial elements of H_2.

There is a great amount of flexibility in the choice of Σ, so Theorem 5.9 gives a lot of isospectral pairs of graphs.

(5.11)   What is less clear is how to ensure that the pairs of isospectral graphs obtained by this method are not isomorphic to each other. This is usually the challenging part! Also, in the work of Sunada for Riemannian manifolds this was the difficult part. For example, in order to use his method to construct isospectral non-isomorphic Riemann surfaces, Sunada used the deep result of Margulis saying that for a non-arithmetic lattice Γ in $SL_2(\mathbb{R})$, Γ is of finite index in the commensurability group of Γ (see [Su]). One cannot imitate this for graphs - the analogous situation would be to look at a uniform lattice Δ in $\mathrm{Aut}(X_k)$ where X_k is the k-regular tree. But the commensurability group of such Δ is always dense in $\mathrm{Aut}(X_k)$ - see [BK].

(5.12)   We do not know any good general argument or condition which will ensure that a pair of graphs constructed as in Theorem 5.9 are not isomorphic. But in practice it seems possible to do so by ad-hoc arguments. here is an example:

EXAMPLE: The following is a Sunada pair due to Gerst [Ge]: Let G be the semi-direct product $G = (\mathbb{Z}/8\mathbb{Z})^* \ltimes \mathbb{Z}/8\mathbb{Z}$, $H_1 = ((\mathbb{Z}/8\mathbb{Z})^*, 0) = \{(1,0),(3,0),$ $(5,0),(7,0)\}$ and $H_2 = \{(1,0),(3,4),(5,4),(7,0)\}$. Take $\Sigma = \{(1,1),(1,7),$ $(3,0),(7,0)\}$. (Note that the last two are of order two). For $i = 1,2$, $X_i = X(G/H_i, \Sigma)$ are both graphs of order 8 which one can draw explicitly. Even without doing so we can see that they are not isomorphic. Indeed, $(3,0)$ and $(7,0)$ in Σ have conjugates in H_i, thus (see Remark 5.3(b)) they gives rise to loops on X_i. The total number of loops in both graphs is the same. (This must be the case by the isospectrality, or if one prefers by the Sunada condition). On the other hand, the distribution of the loops is not determined by the spectrum. So, in X_1 the coset $eH_1 = (1,0)H_1$ has two loops since $(3,0)$ and $(7,0)$ are both in H_1. We claim that no vertex of X_2 has two loops attached to it. Indeed, the cosets of G/H_2 are represented by $\{(1,i) \mid i \in \mathbb{Z}/8\mathbb{Z}\}$, so for $y_i = (1,i)H_2$ to have two loops we need

(a) $(1,i)^{-1}(3,0)(1,i) \in H_2$ and

(b) $(1,i)^{-1}(7,0)(1,i) \in H_2$.

    This means

(a) $(3, -3i + i) \in H_2$ and

(b) $(7, -7i + i) \in H_2$.

The solution of (a) is $i = 2$ or 6, and the solution of (b) is $i = 0$ or 4. Thus there is no vertex with two loops attached to it.

Incidently, neither X_1 nor X_2 are regular graphs: X_1 has vertices of degree 6 (e.g., $(1,0)H_1$) and of degree 4 and X_1 has vertices of degree 5 and of degree 4.

(5.13)   It is tempting to conjecture that all pairs of isospectral graphs are Sunada's pairs. Brooks [Br] showed that this is actually not the case. But he showed that some of the classical constructions of isospectral graphs are special cases of the Sunada way.

REFERENCES

[Al] D.J. Aldous, *Random walks on finite groups and rapidly mixing Markov chains*, In Séminaire de Probabilités 17, Lecture Notes in Math. 298, Springer 1983, 243–297.

[AR] N. Alon, Y. Roichman, *Random Cayley Graphs and Expanders*, Random Structures and Algorithms, 5(1994) 271–284.

[AH] Z. Arad, M. Herzog, *Products of Conjugacy Classes in Groups*, Lecture Notes in Math. 1112, Springer-Verlag 1985.

[AFM] A. Arad, E. Fisman, M. Myzuchuk, *Order evaluation of products of subsets in finite groups and its applications*, Bar-Ilan University, preprint.

[Bab] L. Babai, *Local expansion of vertex transitive graphs and random generation of finite groups*, Proc. 23rd ACM STOC (1991) 164–174.

[BHKLS] L. Babai, G. Hetyei, W. M. Kantor, A. Lubotzky, A. Seress, *On the diameter of finite groups*, 31 IEEE Symp. on Foundations of Computer Science (FOCS 1990), 857-865.

[Ba] R. Bacher, *Valeur propre minimale du laplacian de Coxeter pour le groupe symétrique*, J. of Alg. **167**(1994) 460–472.

[BDH] R. Bacher, P. de la Harpe, *Exact values of Kazhdan constants for some finite groups*, J. of Alg., **163**(1994) 495–515.

[BK] H. Bass, R. Kulkarni, *Uniform tree lattices*, J. of A.M.S., **3**(1990), 843–902.

[Bo] B. Bollobás, *Graph Theory*, Springer-Verlag 1979.

[Br] R. Brooks, in preparation.

[CDGT] D. Cvetkovic, M. Doob, I. Gutman and A. Tarĝasev, *Recent Results in the Theory of Graph Spectra*, Annals of Discrete Math. 36, North-Holland, Amsterdam 1988.

[CDS] D.M. Cvetkovic, M. Doob and H. Sachs, *Spectra of Graphs*, Academic Press, N.Y. 1979.

[CF] J.W.S. Cassels, A. Fröhlich (eds.) *Algebraic Number Theory*, Academic Press, London 1967.

[D] P. Diaconis, *Group Representation in Probability and Statistics*, Inst. of Math. Statistics, Howard, California, 1988.

[Di] J. Dixon, *The probability of generating the symmetric group*, Math. Z. **110**(1969) 199–205.

[DS] P. Diaconis, M. Shahshahani, *Generating a random permutation with random transpositions*, Z. Wahrscheinlichkeitstheorie verw. Gebriete **57**(1981), 159–179.

[FL] S. Fomin, N. Lulov, *On the number of rim hook tableaux*, preprint.

[FJRST] J. Friedman, A. Joux, Y. Roichman, J. Stern and J.P. Tillich, *Most regular graphs are quickly r-transitive*, in preparation.

[FOW] L. Flatto, A.M. Odlyzko and D.B. Wales, *Random shuffles and group representations*, Ann. of Probability **33**(1985) 154–178.

[Ga] F. Gassmann, *Bemekungen zur varstenhenden Arbeit von Harwitz*, Math. Zeitschr. **25**(1926) 665–675.

[Ge] I. Gerst, *On the theory of n-th power residues and a conjecture of Kronecker*, Acta. Arith. **17**(1970), 121–139.

[GKS] R. M. Guralnick, W. M. Kantor and J. Saxl, *The probability of generating a classical group*, Comm. in Alg. **22**(1994) 1395–1402.

[Gl1] D. Gluck, *Character value estimates for groups of Lie type*, Pacific J. **150**(1991) 279–307.

[Gl2] D. Gluck, *Character value estimates for non-semisimple elements*, J. Alg. **155**(1993) 221–237.

[Gr] Y. Greenberg, Ph.D. Thesis, The Hebrew University 1995.

[Gs] J.L. Gross, *Every connected regular graph of even degree is a Schreier coset graph*, J. of Comb. Th. Ser. B. **22**(1977) 227–232.

[GWW] C. Gordon, D. Webb and S. Wolpert, *One cannot hear the shape of a drum*, Bull. A.M.S. **27**(1992) 134–138.

[Hi] M. Hildebrand, *Generating random elements in $SL_n(\mathbb{F}_q)$ by random transvections*, J. of Alg. Comb. 1(1992) 133–150.

[Ke] H. Kesten, *Symmetric random walks on groups*, Trans. AMS **92**(1959), 336–354.

[KL] W.M. Kantor, A. Lubotzky, *The probability of generating a finite classical group*, Geometric Dedicata **36**(1990), 67–87.

[Le] F.T. Leighton, *Finite common coverings of graphs*, J. Comb. Th. B**33**(1982) 231–238.

[LS] M. Liebeck, A. Shalev, *The probability of generating a finite simple group*, Geom. Ded. to appear.

[Lu] A. Lubotzky, *Discrete Groups, Expanding Graphs and Invariant Measures*, Progress in Mathematics Vol. 125, Birkhäuser 1994.

[Lul] N. Lulov, *Generating a random permutation by random involution*, in preparation.

[LW] A. Lubotzky, B. Weiss, *Groups and expanders*, in: "Expanding graphs"

95-109, DIMACS series Vol. 10, American Math. Soc. 1993 (Ed: J. Friedman).

[P1] U. Porod, *The cut-off phenomenon for random reflections*, preprint.

[P2] U. Porod, *L^2 lower bounds for special class of random walks*, preprint.

[R1] Y. Roichman, *Cayley graphs of the symmetric groups*, Ph.D. Thesis, The Hebrew University of Jerusalem 1994 (in Hebrew).

[R2] Y. Roichman, *Upper bounds for characters of the symmetric groups*, preprint.

[R3] Y. Roichman, *Random walks on the alternating groups with conjugacy classes*, preprint.

[R4] Y. Roichman, *Expansion properties of Cayley graphs of the symmetric group with respect to cycles*, preprint.

[Sc] G. Schechtman, *Levy type inequality for a class of finite metric spaces*, in "Martingale theory in harmonic analysis and Banach spaces" (Eds: J.A. Chao and W.A. Woyzynski) Lect. Notes in Math. **939**(1982) 211–215, Springer-Verlag.

[Se] J. J. Serre, <u>Trees</u>, Springer-Verlag, Berlin-Heidelberg-New York, 1980.

[Su] T. Sunada, *Riemannian coverings and isospectral manifolds*, Ann. of Math. **121**(1985) 169–186.

Construction and Classification of Combinatorial Designs

EDWARD SPENCE

Department of Mathematics
University of Glasgow
Glasgow, G12 8QW
SCOTLAND

§1 Introduction

One important area of research in Combinatorial Mathematics is that of the existence, construction and enumeration of designs of various different sorts. Given the general conditions defining a design it is often the case that very simple necessary conditions for its existence can be derived in terms of the so-called parameters of the design, but by and large it is very difficult to prove or disprove that the conditions obtained are also sufficient. Thus, even if necessary conditions are obtained we are generally left with either an immediate non-existence result or the problem of attempting to establish the existence by an explicit construction of the design. Of course, if the design concerned does not actually exist then this can often be a lengthy process. The first example that springs to mind is a projective plane of order 10, where the necessary conditions obtained by the Bruck-Ryser-Chowla Theorem do not disprove its existence, and yet the plane does not exist [16].

If the parameter sets of the designs are 'small' then it has often been the case that direct computer-free methods have led to a construction, and in some cases a complete classification has been achieved. In certain sporadic instances with 'large' parameter values it is also possible to construct and classify the corresponding designs without using a computer, but it is generally true that the larger the parameter set the more difficult the problem of determining all non-isomorphic designs with that given set of parameters. In many cases the problem is beyond the capabilities of the mathematician on his own and if he is seriously interested in a full classification he has to call upon the assistance of the computer. This is nothing new since many mathematicians in general, and combinatorialists in particular, have been using computers as an aid since they were invented. What is new, however, is my own personal involvement in this direction. Some nine years ago I wished to determine the automorphism group of a certain symmetric design (a 2-$(36, 15, 6)$ design in fact) and at that time I had no idea how to begin. I had no knowledge as to how to use a computer, let alone how it might be used to help solve the problem . Then I discovered at first hand that Frans Bussemaker of Eindhoven University of Technology had a computer program that could give the answer in a few seconds. From that moment on I was converted, and now the computer is an indispensable aid to my research into the classification of combinatorial designs. However, since the area comprising designs in general is so vast we shall, in the subsequent sections, confine ourselves to considering only regular two-graphs, strongly regular graphs and 2-designs, where, in the main, the 2-designs will be *symmetric*. We shall give a survey of some of the methods that have been applied to attack the problem of classifying some of these designs using the computer. Of course, we have to be a little bit clever for the following reason (among others). Since any permutation of the points and any permutation of the blocks of a design give an isomorphic copy of the design, there will be, even for relatively small numbers of points and blocks, a vast number of branches of any search tree that have to be pruned to give an efficient algorithm. Without such even the fastest of computers would be hard pushed to complete some searches in a time measured in years, rather that in hours or even days.

In order to give the reader an idea of the present state of knowledge concerning the numbers of non-isomorphic symmetric 2-(v, k, λ) designs for values of $v <= 45$, we include the following table with the relevant data. We exclude finite projective planes ($\lambda = 1$), for it is known that they are unique for k a prime power less than 9. Also excluded is the parameter set $(43, 21, 10)$, for which it is known that there are at least 2. Some of the entries were obtained from the tables in [19] and others have been obtained as a result of some of the author's work that is described in the following sections.

v	k	λ	No. of Designs	v	k	λ	No. of Designs
11	5	2	1	45	12	3	≥ 3752
16	6	2	3	40	13	4	≥ 1465
15	7	3	5	27	13	6	208310
37	9	2	4	36	15	6	≥ 25634
25	9	3	78	31	15	7	≥ 1266891
19	9	4	6	41	16	6	≥ 112000
31	10	3	151	35	17	8	≥ 108131
23	11	5	1103	39	19	9	≥ 38

Finally, lest the reader gets the impression that the author believes the computer to be the answer to every problem, we mention the construction of a symmetric 2-(160,54,18) design, where the parameters are so large that they effectively render the computer useless.

§2 Symmetric 2-designs

A 2-(v, k, λ) design, $(0 < \lambda < k < v)$, is a pair $(\mathcal{P}, \mathcal{B})$, where $\mathcal{B}$ is a collections of k-subsets (the *blocks*) from a v-set $\mathcal{P}$ (the *points*) such that each pair of points of $\mathcal{P}$ is contained in exactly λ blocks of $\mathcal{B}$. A straightforward counting argument shows that each point is in the same number r say, of blocks and that if b denotes the number of blocks then $bk = vr$ and $r(k-1) = \lambda(v-1)$. Fisher's inequality [10] shows that $b \geq v$ and when equality is achieved the 2-design is said to be *symmetric*. Perhaps the best-known existence result concerning symmetric 2-designs is the following ([10], Theorem 10.3.1).

Theorem 2.1 (Bruck-Ryser-Chowla) *Suppose that a symmetric 2-(v, k, λ) design exists. Then*
1. *If v is odd, the Diophantine equation $z^2 = (k - \lambda)x^2 + (-1)^{(v-1)/2}\lambda y^2$ has a solution in integers x, y, z not all zero.*
2. *If v is even, $k - \lambda$ is a square.*

To date there is known only one parameter set, namely $(v, k, \lambda) \equiv (111, 11, 1)$, that satisfies the conditions of this theorem and for which a corresponding design does not exist [16]. For possible designs with 'small' parameter sets not ruled out, for example $(7, 3, 1)$, $(13, 4, 1)$ and $(11, 5, 2)$, a simple direct attempt at construction brings immediate rewards. Not only are the designs easy to construct but the very process yields their uniqueness, up to *isomorphism*. This method, however, is clearly not generally feasible for designs with larger block size. In that case a well-tried technique for design construction is to assume the existence of a non-trivial automorphism (usually of a fairly high order) and to utilise its action on the points and blocks of the hypothetical design. In some cases it is possible to do the analysis by hand and indeed many of the first known designs with a given parameter set were constructed by this method. However, in the absence of such an automorphism, or even with one of small order, the task is much more difficult.

In what follows we shall mainly consider *symmetric* 2-designs and before we give a brief description of the ideas involved we first quote the following results from [17] which will prove useful. As above, the symmetric 2-design will have points set $\mathcal{P}$ and block set $\mathcal{B}$.

Theorem 2.2 ([17] Theorem 3.1) *An automorphism σ of a symmetric 2-design fixes equally many points and blocks.*

Theorem 2.3 ([17] Theorem 3.20) *Suppose that a symmetric* 2-(v, k, λ) *design has an automorphism* σ *of prime order* q. *Let* f *be the number of fixed points of* σ *and* $w + f$ *the number of orbits of* σ *on* $\mathcal{P}$. *Then*

1. *Either* $k - \lambda$ *is a square or* $w + f$ *is odd.*
2. *Suppose that for some prime* p *dividing* $k - \lambda$ *there is an integer* j *such that* $p^j \equiv -1 \pmod{q}$. *Then* w *is even and* f *is odd.*

Thus an automorphism σ of order q partitions both $\mathcal{P}$ and $\mathcal{B}$ into $f + w$ orbits of which f have size 1 (corresponding to the fixed points and fixed blocks) and w have size q, so that $v = f + qw$. The next step is to choose a representative from each of the $f + w$ orbits of blocks and form the $(f + w) \times (f + w)$ orbit matrix $Q_\sigma = [Q_{ij}]$, where Q_{ij} is the number of points of the ith point orbit incident with the jth block representative. It is easy to see that Q_σ satisfies the conditions

$$\sum_{\ell=1}^{f+w} Q_{i\ell} = k, \qquad \sum_{\ell=1}^{f+w} Q_{i\ell} Q_{j\ell} / s_\ell = \lambda, \tag{2.1}$$

where s_i is the size of the ith orbit $(1 \leq i, j \leq f + w, \ i \neq j)$.

The matrix Q_σ is of smaller dimensions than the incidence matrix of the design, and although its entries may range from 0 to q it is generally easier to solve (2.1) as a preliminary step, or to show that it has no solution, than to attempt to construct an incidence matrix directly. In some cases an orbit matrix can easily be found, but if a *complete* classification of the designs with an automorphism of order q is required, it is often the case that a computer is needed. Of course, not every solution of (2.1) necessarily arises as an orbit matrix of some design.

Let us assume that a solution Q_σ of (2.1) has indeed been found, and has the form

$$Q_\sigma = \begin{bmatrix} F & U \\ V & W \end{bmatrix},$$

say, where F and W are square matrices of orders f and w, respectively, and F is the incidence matrix of the fixed points and fixed blocks. The non-zero entries of U are q and 1, those of V are 1 and W is an integral matrix with entries in the range $0 \ldots q$. An attempt is then made to find a new matrix P, say whose entries P_{ij} are $q \times q$ circulant $(0, 1)$ matrices with row sums W_{ij}, such that the matrix

$$\begin{bmatrix} F & \mathbf{j}^t \otimes U/q \\ \mathbf{j} \otimes V & P \end{bmatrix}, \tag{2.2}$$

of size $f + qw = v$, is an incidence matrix of a 2-(v, k, λ) design. Here $\mathbf{j}$ is the all-one vector of size q.

We now illustrate the ideas involved by referring to one particular example. In [1] the authors produced the first known symmetric 2-(41, 16, 6) design by assuming it had an automorphism of order 5. They then showed that this automorphism had to have precisely one fixed point (and therefore precisely one fixed block). A (computer) search for orbit matrices satisfying (2.1) produced a total of 15 and one of these, having a further automorphism of order 3, was chosen. Using it the authors were able to construct a design quite simply, and since their sole aim was to establish the existence of one such a design, they attempted no further investigation. A full classification, however, would require a detailed examination of *all* the orbit matrices found and in this respect their list of 15 was somewhat deficient. It turned out that they had omitted from their findings a further 3 orbit matrices (up to what might loosely be described as 'isomorphism and duality'); the correct number is 18 [30]. Only one of these orbit matrices failed to produce a design (2.2) when a complete classification was effected using a computer and a backtracking search.

The only other possibilities for prime orders of (non-trivial) 2-(41, 16, 6) designs are 2 and 3 [1]. In [30] it was shown that an automorphism of order 3 had to have 5 or 11 fixed points. In the former case there were 960 solutions to (2.1), all of which, except one, produced designs, while in the latter, the orbit matrix was unique. That this orbit matrix was unique was easily established using a computer-free argument, but because there were so many designs arising from it, to determine the exact number (3076) required the use of the computer once again.

It is clear that the methods outlined can work only in the case it is known that a non-trivial automorphism exists. Even then, although the method in theory will work, in practice it is often a different proposition. It might seem an easy task to determine, for example, all 2-(40, 13, 4) designs having an involution, but this is far from the case. There are several possibilities for the numbers of fixed points for such, and some of these give rise to very many solutions to (2.1). For example, in the case of fixed-point-free involutions, we have found that there are at least 20603 non-isomorphic solutions (there are certainly many more), and these orbit matrices give rise to 1003 2-(40, 13, 4) designs [35]. Thus the majority of these orbit matrices do not yield designs. But despite spending a great deal of time on the problem, we have been unable to write a really efficient program that will weed out at an early stage those orbit matrices that cannot be extended. On the other hand, if the involution has 10 fixed points, there are relatively few orbit matrices (7 in fact) that yield 78 non-isomorphic designs [35]. By way of a contrast to the above, Čepulić in [5] considered 2-(40, 13, 4) designs having an automorphism of order 5 and he showed that there were precisely 4 orbit matrices yielding 13 non-isomorphic designs. It appears that, in general, the larger the order of the automorphism or the greater the number of fixed points, the greater the likelihood of there being few orbit matrices, with a corresponding reduction in the complexity of the computer work involved.

In a previous paper [27] the author made a partial classification of the strongly regular graphs with parameters $(40, 12, 2, 4)$ and using these he was able to construct several hundred 2-$(40, 13, 4)$ designs, many with fairly large automorphism groups and many with a trivial group. On the basis of this and his further experience in attempting to determine completely those designs that have an automorphism of order 2 or 3 (the only other possibilities for an automorphism of prime order), he feels that a total classification is many years off. Combining the result of Čepulić and the findings of the author, however, establishes the existence of at least 1465 pairwise non-isomorphic 2-$(40, 13, 4)$ designs.

Similar comments apply to the 2-$(36, 15, 6)$ case. As we shall see in §6 there are many thousands of such designs that arise from regular two-graphs on 36 vertices. All these designs have a polarity with no absolute points and so have a symmetric incidence matrix. The fact that the computer search for these regular two-graphs is not yet complete surely indicates that, if the symmetry is not assumed, then a general classification is utterly hopeless.

Other authors have used the above ideas to generate symmetric 2-designs. For example, Tonchev [38] and Mathon [18], as well as the author [28], have all made contributions to the construction of 2-$(31, 10, 3)$ designs with a non-trivial automorphism culminating in their complete classification, and Tonchev [36], [37] applied similar techniques in his search for 2-$(27, 13, 6)$ designs.

Since the existence of even a non-trivial automorphism does not guarantee quick and easy classification, the problem of *complete* classification involving also designs with trivial group might at this stage seem completely hopeless. We discuss this in §4.

§3 (v, k, λ) Graphs and Designs

A (v, k, λ) graph G is a strongly regular graph with $\lambda = \mu$. Thus its $(0, 1)$ adjacency matrix A satisfies

$$A^2 = (k - \lambda)I + \lambda J, \quad Ak = kJ, \quad A^t = A.$$

These equations can also be interpreted as saying that A is the point-block incidence matrix of a symmetric 2-(v, k, λ) design $D(G)$ say, having a polarity with no absolute points, cf [22]. It is possible for non-isomorphic (v, k, λ) graphs to produce isomorphic 2-(v, k, λ) designs, as happens in the $(16, 6, 2)$ case. On the other hand, it was shown in [8] that if G_1 and G_2 are two non-isomorphic (v, k, λ) graphs for which one of $\mathrm{Aut}(G_1)$, $\mathrm{Aut}(G_2)$ has odd order, then the corresponding designs $D(G_1)$, $D(G_2)$ are non-isomorphic. In [3], [9] the following question was posed:

Is $\mathrm{Aut}(G) = \mathrm{Aut}(D(G))$ if $\mathrm{Aut}(G)$ has odd order?

There it was proved that for $v < 63$ the answer is in the affirmative except for three graphs, one of which was a $(36, 15, 6)$ graph, the other two being $(45, 12, 3)$ graphs. The investigation of the $(45, 12, 3)$ case led to the discovery of a large number of non-isomorphic 2-$(45, 12, 3)$ designs, many with a *trivial* automorphism group [20]. Before giving a brief description of the methods used we first recall that a square matrix M is *anti-cyclic* if

$$M_{i,\,j+1} = M_{i+1,\,j} \qquad (1 \le i, j \le n),$$

the suffices being taken mod n. The next result is quoted from [3].

Theorem 3.1 ([3], Theorem 4) *Let A be the adjacency matrix of a (v, k, λ) graph G. Suppose $\mathrm{Aut}(G) \ne \mathrm{Aut}(D(G))$, and $\mathrm{Aut}(G)$ has odd order. Then there exist a prime $p > 2$, an integer $m < v/p$, and a partitioning*

$$A = \begin{bmatrix} A_{00} & A_{01} & \cdots & A_{0m} \\ A_{10} & A_{11} & \cdots & A_{1m} \\ \vdots & \vdots & \cdots & \vdots \\ A_{m0} & A_{m1} & \cdots & A_{mm} \end{bmatrix},$$

such that
    (i) $p \le (k\sqrt{(k - \lambda)} - k + \lambda)/\lambda$,
    (ii) $(p - 1)(m - 1) \ge 2(k - \lambda)$,
    (iii) $A_{ii} = 0$ for $i = 1, 2, \ldots, m$,
    (iv) $A_{ij} = A_{ji}$ is anti-cyclic of size $p \times p$, for $i, j = 1, 2, \ldots, m$,
    (v) $A_{i0} = A_{0i}^t$ consists of p identical rows for $i = 1, 2, \ldots, m$,
    (vi) $m(k - \lambda)$ is even if $p = 3$.

Conditions (iv) and (v) arise from the fact that $D(G)$ necessarily has an automorphism whose action is described by $QAQ = A$, where

$$Q = \mathrm{diag}\{1, 1, \ldots, 1, P, P, \ldots, P\},$$

the diagonal blocks P being cyclic permutation matrices of size p.

In [3] each of the four parameter sets $(35, 18, 9)$, $(36, 15, 6)$, $(36, 21, 12)$ and $(40, 27, 18)$ was considered and there it was shown that there is only one graph with trivial group and one of these sets of parameters, namely $(36, 15, 6)$, that has a non-trivial automorphism when considered as a design.

In fact the result is also true if we replace 'trivial group' with 'automorphism group of odd order'.

In the $(45, 12, 3)$ case it was shown that the only two possibilities were either (a) $p = 3$, $m = 12$, or (b) $p = 3$, $m = 14$. In the respective cases it was easy to show, without the use of the computer, that the adjacency matrices have to take the forms:

$$
J_3 \otimes P_6 \begin{bmatrix}
r_1 & r_2 & r_3 & r_1 & r_2 & r_3 & r_1 & r_2 & r_3 \\
r_1 & r_2 & r_3 & r_3 & r_1 & r_2 & r_2 & r_3 & r_1 \\
r_1 & r_2 & r_3 & r_2 & r_3 & r_1 & r_3 & r_1 & r_2 \\
1 & 1 & 1 & 1 & 1 & 1 & 1 & 1 & 1 \\
1 & 1 & 1 & 1 & 1 & 1 & 1 & 1 & 1 \\
1 & 1 & 1 & 1 & 1 & 1 & 1 & 1 & 1 \\
0 & 0 & 0 & 1 & 1 & 1 & 1 & 1 & 1 \\
 & 0 & 0 & 1 & 1 & 1 & 1 & 1 & 1 \\
 & & 0 & 1 & 1 & 1 & 1 & 1 & 1 \\
 & & & 0 & 0 & 0 & 1 & 1 & 1 \\
 & & & & 0 & 0 & 1 & 1 & 1 \\
 & & & & & 0 & 1 & 1 & 1 \\
 & & & & & & 0 & 0 & 0 \\
 & & & & & & & 0 & 0 \\
 & & & & & & & & 0
\end{bmatrix},
$$

$$
J_3 \otimes P_6 \begin{bmatrix}
r_1 & r_2 & r_3 & r_2 & r_3 & r_1 & r_3 & r_1 & r_2 \\
1 & 1 & 1 & 1 & 1 & 1 & 1 & 1 & 1 \\
1 & 1 & 1 & 1 & 1 & 1 & 1 & 1 & 1 \\
1 & 1 & 1 & 1 & 1 & 1 & 1 & 1 & 1 \\
1 & 1 & 1 & 1 & 1 & 1 & 1 & 1 & 1 \\
1 & 1 & 1 & 1 & 1 & 1 & 1 & 1 & 1 \\
0 & 0 & 0 & 0 & 2 & 1 & 0 & 1 & 2 \\
 & 0 & 0 & 1 & 0 & 2 & 2 & 0 & 1 \\
 & & 0 & 2 & 1 & 0 & 1 & 2 & 0 \\
 & & & 0 & 0 & 0 & 0 & 2 & 1 \\
 & & & & 0 & 0 & 1 & 0 & 2 \\
 & & & & & 0 & 2 & 1 & 0 \\
 & & & & & & 0 & 0 & 0 \\
 & & & & & & & 0 & 0 \\
 & & & & & & & & 0
\end{bmatrix},
$$

(a)   (b)

where

$$
r_1 = \begin{bmatrix} 1 & 1 & 1 \\ 0 & 0 & 0 \\ 0 & 0 & 0 \end{bmatrix}, \quad
r_2 = \begin{bmatrix} 0 & 0 & 0 \\ 1 & 1 & 1 \\ 0 & 0 & 0 \end{bmatrix}, \quad
r_3 = \begin{bmatrix} 0 & 0 & 0 \\ 0 & 0 & 0 \\ 1 & 1 & 1 \end{bmatrix},
$$

P_6 is a permutation matrix of order 6 representing a symmetric derangement, J_3 is the all-one matrix of order 3 and an entry 0, 1, 2 indicates an anticyclic sub-matrix of order 3 with row sums 0, 1, 2, respectively.

It turned out to be a relatively simple task to determine all $(45, 12, 3)$ graphs arising from these tactical configurations. In the first case there were 16 and in the second 29. As a result we were able to identify the two $(45, 12, 3)$ graphs whose automorphism group of odd order was a proper subgroup of the automorphism group of the corresponding design.

What was more challenging was a consequence of the following observation. If in each of the above 45 graphs the symmetric permutation matrix P_6 is replaced by an arbitrary permutation matrix, the symmetry of the corresponding adjacency matrix may be destroyed, but the resulting matrix (also denoted by A) still satisfies the condition $AA^t = 9I + 3J$. Thus, this new matrix is the incidence matrix of a 2-$(45, 12, 3)$ *design*. Also, it was apparent that removing the condition that the sub-matrices of size 3 had to be anti-cyclic would probably result in the construction of further non-isomorphic graphs and designs. By utilising the automorphisms of the matrices (a) and (b) we were able to determine completely the numbers of non-isomorphic 2-$(45, 12, 3)$ designs obtained in this way. In total there are 3745 such designs, of which 1136 have trivial automorphism group, the remainder having automorphisms of order 2, 3 or 5. It is doubtful if so many designs with trivial automorphism group could be found using any other method.

It was possible to determine all the designs with automorphisms of order 5 or 11 using the strategy of §2 (the only other possibilities for automorphisms of prime order) but, on the basis of the numbers produced above, it would seem likely that, as in the $(40, 13, 4)$ case, a complete classification of designs with these parameters is still a long way off. These results are detailed more fully in [20].

§4 Symmetric Designs with Trivial Group

In the early eighties Denniston [6] successfully used the computer to classify all 2-$(25, 9, 3)$ designs and at that time this was certainly a great achievement. Moreover, Ito et al [11] seemed to have found all Hadamard matrices of order 24 and so have enumerated all 2-$(23, 11, 5)$ designs. However, it turned out that they had unfortunately missed one, a fact discovered by Kimura [13] and later verified independently by the author [33]. The next two 'smallest' symmetric designs for which a complete classification was at that time unknown were those with parameters 2-$(27, 13, 6)$ and 2-$(31, 10, 3)$. As we saw above, if only designs with a non-trivial automorphism group are considered, the latter case is completely worked out. It had seemed unlikely that much progress could be made in the general case because of the demands on memory and disc space, but by using a simple algorithm that required the search to be split into several parts the author was able to identify *all* 2-$(31, 10, 3)$ designs [29].

Let A be a $v \times b$ matrix with entries 0 or 1 and suppose that for a positive integer n, S_n denotes the group of permutation matrices of order n. Corresponding to A we have the binary integer, which we denote by $\nu(A)$, obtained by concatenating the rows of A. We then define the *Standard form* of A to be the the $v \times b$ matrix B given by

$$\nu(B) = \max \{\nu(PAQ) : P \in S_v, Q \in S_b\}.$$

Further, for each integer r $(1 \leq r \leq v)$, let A_r denote the $r \times b$ matrix comprising the first r rows of A. A simple but extremely useful observation is that if A is in standard form then so also is A_r $(1 \leq r \leq v)$.

We shall now assume that A is an incidence matrix in standard form of a symmetric 2-(v, k, λ) design, so that its first two rows and columns are as shown, the first two rows and columns having λ 1's in common.

$$A = \begin{bmatrix}
1 & 1 & \cdots & 1 & 1 & \cdots & 1 & 0 & \cdots & 0 & 0 & \cdots & 0 \\
1 & 1 & \cdots & 1 & 0 & \cdots & 0 & 1 & \cdots & 1 & 0 & \cdots & 0 \\
\vdots & \vdots & & & & & & & & & & & \\
1 & 1 & & & & & & & & & & & \\
1 & 0 & & & & & & & & & & & \\
\vdots & \vdots & & & & & & & & & & & \\
1 & 0 & & & & & & & & & & & \\
0 & 1 & & & & & & & & & & & \\
\vdots & \vdots & & & & & & & & & & & \\
0 & 1 & & & & & & & & & & & \\
0 & 0 & & & & & & & & & & & \\
\vdots & \vdots & & & & & & & & & & & \\
0 & 0 & & & & & & & & & & &
\end{bmatrix}.$$

The procedure is as follows. Suppose that A_{r-1} $(r \geq 3)$ has already been constructed. In building the rth row of A we have to satisfy the conditions that its weight should be k and that its inner products with rows 1 to $r-1$ are all λ. Let $WT(r,s)$ denote the number of 1's in the first s positions of row r, (the weight of the rth row as far as the sth column), and let $IP(r,s,i)$ $(1 \leq i < r)$ denote the inner product of the ith and rth rows as far as the sth column. Obviously

$$k - v + s \leq WT(r,s) \leq k, \qquad IP(r,s,i) \leq \lambda \ (1 \leq i < r \leq v),$$

with equality in both cases when $s = v$. The first of these two inequalities cannot really be bettered, but the second, being highly inefficient, needs to be improved.

Lemma 4.1 *Suppose that $1 \leq i < r \leq v$. Then, writing n for $k - \lambda$ and $\overline{\lambda} = v - 2k + \lambda$,*

$$\max\{WT(r,s), WT(i,s)\} - n \leq IP(r,s,i) \leq \overline{\lambda} - s + WT(i,s) + WT(r,s).$$

Proof. Since the proof is a straightforward counting argument, it is omitted. □

Of course the inequality on the right-hand side is an improvement only if $v - 2k - s + WT(i,s) + WT(r,s) < 0$.

There are thus four conditions that need to be satisfied at position (r,s) of the rth row of A.

(1) $IP(r,s,i) \geq WT(i,s) - (k - \lambda)$.

(2) $IP(r, s, i) \geq WT(r, s) - (k - \lambda)$.
(3) $IP(r, s, i) \leq \lambda$.
(4) Let $N_s = v - 2k - s + WT(i, s) + WT(r, s)$. Then $IP(r, s, i) \leq \lambda + N_s$.

We therefore suppose that these conditions are all satisfied at position $(r, s - 1)$ of the rth row of A and determine which of them require to be tested at position (r, s), since some of them may well be redundant. The following results are easily established.

$A[r, s]$	$A[i, s]$	Condition
1	1	3
1	0	2
0	1	1
0	0	4

Using these conditions we then attempt to construct rows 3 to v in turn, entries one at a time, so that by the time the rth row is completed A_r is in standard form. It is imperative that this be the case, at least for small values of r, for otherwise we may well go down a path of the search tree that has already been traced. There are some simple and obvious tests that can be applied, as each row is being constructed, that will reject a choice for the (r, s) entry when it is clear that with this choice the sub-matrix A_r will not be in standard form. For example, the $(r - 1)$th row of A must be lexicographically greater than the rth row, with a similar observation for the columns. However, even with these tests it may still happen that A_r is not in standard form. For this reason, we applied the procedure DESIGNPERMUTATIONSTANDARD of Frans Bussemaker to test whether or not this was the case; if not, the rth row was rejected and an attempt was made to construct another one using a backtracking procedure. This test was extremely efficient for small values of r but it became less useful for larger values. One of the reasons for this was that very often there would be no way of constructing an $(r + 1)$th row compatible with A_r so applying the test in this case would be unnecessary. In the two cases where we applied the algorithm as it is described here, namely $(v, k, \lambda) \equiv (25, 9, 3)$ and $(31, 10, 3)$, we chose a threshold value of $r = k$. It was then relatively easy to construct A_k. In the respective cases above we found 36 and 960 possibilities (in standard form) thus confirming previously found figures for the numbers of 2-$(9, 3, 2)$ and 2-$(10, 3, 2)$ designs [19]. However, applying the same strategy in the $(31, 10, 3)$ case to finding $A_{k+1}, \ldots, A_v$ was clearly not viable because of the time it would require. We had to proceed differently. Our attempt to extend A_k to A was done in two stages, the first stop being $A_{k+(k-\lambda)}$. Observing that the rows numbered $k + 1$ to

$k + (k - \lambda)$ must begin $0, 1, \ldots$ and must be in lexicographical order, we determined, for each A_k, all $(0, 1)$ rows $(0, 1, \ldots)$ of length v and weight k that were compatible with A_k, in other words, whose inner products with the rows of A_k were all λ. Let us denote these rows by

$$\rho_1 > \rho_2 > \ldots > \rho_m,$$

where the ordering is lexicographical. Now construct a graph Γ_k say, whose vertices are these rows and where adjacency is defined by

$$(\rho_1, \rho_j) \text{ is an edge if and only if } \rho_i \, \rho_j^t = \lambda.$$

A clique of Γ_k of size $k - \lambda$, say $\{\rho_{i_1} > \rho_{i_2} > \ldots > \rho_{i_{k-\lambda}}\}$, gives a completion of A_k to $A_{2k-\lambda}$, which may or may not be in standard form. It is therefore necessary to determine all cliques of Γ_k of size $k - \lambda$, use them to obtain $A_{2k-\lambda}$ and then to test whether or not this matrix is in standard form. Since an efficient clique-finding procedure was not difficult to write, and perhaps more importantly, the number m of vertices of Γ_k never exceeded 800, it turned out that this method of generating rows $k + 1$ to $2k - \lambda$ was far superior to that of finding rows 1 to k.

Now that we have reached the stage of constructing all possible $A_{2k-\lambda}$ that can be embedded in an incidence matrix A in standard form, there appear to be several options open to us. The first might naturally be to use a similar procedure to the one just described. Since rows $2k - \lambda + 1$ to v of A must begin with two zeros we should find all $(0, 1)$ rows $(0, 0, \ldots)$ of length v and weight k that are compatible with $A_{2k-\lambda}$, construct a graph with these rows as vertices and then determine all cliques of size $v - 2k + \lambda$. However, there were in general too many cliques to make this worthwhile. The second would be to revert to the original method of completing each row in turn, ensuring as far as possible that the resulting matrix will be in standard form without actually using Bussemaker's procedure (until the ultimate stage). Unfortunately this also was very inefficient and a different method had to be found. In practice we discovered that *transposing* $A_{2k-\lambda}$ and then completing its rows numbered 3 to v was several times faster.

Our first implementation of the algorithm just described was written in Pascal for a 33 Mhz 486 AT computer, under MSDOS, and when applied to the $(31, 10, 3)$ case completed the search in around 50 hours CPU time, finding 151 pairwise non-isomorphic designs, of which 107 have trivial automorphism group. The largest graph Γ_{10} had 812 vertices while the smallest one comprised 292. Overall there were 406112 cliques of size 7, not all of which gave an A_{17} in standard form. As a later check we rewrote it in C and ran it in an HP 710 work station which had the effect of reducing the CPU time to around 10 hours. When applied to the 2-$(25, 9, 3)$ case the search was completed in under 30 minutes.

In the previous section we mentioned symmetric designs with parameter sets $(40, 13, 4)$ and $(45, 12, 3)$ in connection with designs having a non-trivial automorphism. Since both these sets have relatively small values of k and λ it might seem at first sight that the above methods could also be applied to them. However, in each of the cases there are large numbers of possible matrices A_k, and corresponding to those we investigated, the graphs Γ_k have several (many) thousand vertices, thus making clique-finding a memory-consuming as well as time-consuming operation.

§5 Hadamard Matrices

It is well-known that a Hadamard matrix is a square ± 1 matrix whose rows are pairwise orthogonal and that its order must be 1, 2 or a multiple of 4. Indeed it is conjectured that they exist for all possible orders and much work has been done on the subject. What we are interested in here, however, is their complete classification in the case the order is 'small'. At present it seems that 'small' means 'no more than 28', for unpublished investigation by the author suggests that there are thousands of non-isomorphic Hadamard matrices of order 32 and certainly a large number of order 36, as we shall see in the next section.

Corresponding to a Hadamard matrix of order $4m$ there is a class of symmetric designs with parameters $2\text{-}(4m - 1, 2m - 1, m - 1)$. These are obtained by 'normalising' the Hadamard matrix, i.e. multiplying its rows and columns so that the resulting matrix has its first row and column all ones, deleting the first row and column and then replacing the -1's by 0. The same procedure can be applied to the ith row and jth column of the Hadamard matrix, again yielding a $2\text{-}(4m - 1, 2m - 1, m - 1)$ design. Thus a Hadamard matrix of order $4m$ may have associated with it as many as $4m \times 4m$ non-isomorphic *Hadamard designs*, as these $2\text{-}(4m - 1, 2m - 1, m - 1)$ designs are called. Here we describe them by saying that they are *descendants* of the Hadamard matrix. Any one of the descendants can be used to recover the Hadamard matrix, by replacing the zeros by -1's and adjoining a row and column of $+1$'s. Thus to classify Hadamard matrices it is enough to determine them by one of their descendants. We chose the one whose standard form, as described earlier, was greatest; we call this the *standard form* of the Hadamard matrix.

The largest order for which we found a complete classification to be possible was 28 and for this reason we describe the method in relation to Hadamard matrices of that order. Their descendants have parameters $(27, 13, 6)$ and prior to our investigation Tonchev [36], [37] had classified those with automorphisms of order 7 and 13. Kimura [12], [14] had begun his investigation, which he later completed [15]. The reader may be comforted to know that, although Kimura and the author used different methods and worked independently, their final results were the same!

One major difficulty that we found was the sheer size of the problem. It was not possible to enumerate *all* designs with the parameters of a residual design of a 2-$(27, 13, 6)$ design, as was done in the 2-$(31, 10, 3)$ case, since the evidence of an initial investigation indicated that there would many millions of them. Of course, even if it had been possible to classify them all, much of the effort would have been redundant as, by and large, many non-isomorphic residual designs can lead to the same Hadamard matrix.

We attempted, therefore, as far as possible, to determine all those 2-$(13, 6, 5)$ designs that, together with an extra block comprising all 13 points, could be embedded as the first 13 rows of the *standard form* of a greatest descendant. This was done on two stages. First, we generated all A_7 in standard form, as in §3. Then we eliminated those which could not be extended to a greatest descendant. This was achieved by comparing A_7 with the descendants of the ± 1 matrix of size 8×28 obtained from A_7 by adjoining a row and column of $+1$'s. The next step was to extend A_7 to A_{13} using the method involving clique-finding, described in §4. Here the graphs obtained had generally between 1000 and 3000 vertices, with an average number of around 1200, and the number of 6-cliques was sometimes as large as 30000. To test whether each A_{13} found could be embedded in a Hadamard matrix in standard form was too expensive in time, so the only pruning that was performed at this level was to test that A_{13} itself was in standard form.

The penultimate step involved extending A_{13} to A_{27}, and this was done in a similar way to the above. Here the graphs involved had on the average fewer than 100 vertices and the number of 14-cliques seldom exceeded 4.

Finally, we eliminated those 2-$(27, 13, 6)$ designs obtained that were not the greatest descendant of a Hadamard matrix, and this left a total of 487. Thus there are 487 non-isomorphic Hadamard matrices of order 28.

Now that Hadamard matrices of order 28 have been classified it is possible (a) to determine all non-isomorphic skew-Hadamard matrices of order 28 and (b) to enumerate all non-isomorphic 2-$(27, 13, 6)$ designs D, as each of these must be a descendant of one of the above 487 Hadamard matrices. In the case (a) we obtained 54 in total and this was done using two separate methods. We first wrote a procedure, again involving backtracking, specifically to determine the skew-Hadamard matrices and because of the extra information regarding the skew-symmetry it was a relatively short search. Secondly, we examined the equivalence classes of the 487 Hadamard matrices to determine those that contained a Hadamard matrix that could be transformed into one of skew type by permuting its rows and columns and/or multiplying its rows and columns by -1. As far as (b) was concerned, by reducing each descendant to standard form we were able to ascertain that the number of non-isomorphic 2-(27, 13, 6) designs is 208310. The sizes of the automorphism groups range from 1 to 1053, the precise distribution being displayed below.

$|\text{Aut}(D)|$	1	2	3	4	6	9	13	18	27	39	78	1053
No. Designs	206842	736	649	32	38	2	2	3	1	3	1	1

Furthermore, we have determined all pairwise non-isomorphic 2-designs that are *residual* designs of the above, the number found being 2572156.

§6 Regular Two-Graphs

A *two-graph* is a pair (Ω, Δ), where Ω is a finite set (the vertex set) and Δ is a collection of triples $\{\omega_1, \omega_2, \omega_3\}$ of distinct vertices $\omega_1, \omega_2, \omega_3 \in \Omega$, with the property that any 4-subset of Ω contains an even number of triples of Δ. In a *regular* two-graph each pair of vertices is in a constant number of the triples of Δ.

A simple graph (Ω, E) yields a two-graph by the following method. The triples of Δ are precisely the triples of vertices of Ω that carry an odd number of edges. Associated with a two-graph is a class of graphs, called the *switching class* of the two-graph. We refer the reader to [23], [24] for the details as to how this is done in terms of the triples of Δ. For our purposes it is enough to consider a switching class of graphs as an equivalence class under the relation of *switching* on the set of all simple graphs $\Gamma(\Omega, E)$ with vertex set Ω and edge set E.

Definition 6.1 Let A and B be the ∓ 1 adjacency matrices of two graphs Γ_1 and Γ_2 on the same vertex set Ω. Then Γ_1 and Γ_2 are said to be *switching-equivalent* if there exists a diagonal ± 1 matrix D such that $DAD = B$.

Since switching equivalent graphs give rise to the same two-graph, by the construction outlined above, we can identify a two-graph with a switching class of graphs. Further, since switching-equivalent graphs have the same Seidel spectrum (the eigenvalues of the ∓ 1 adjacency matrix), we can define the spectrum of a two-graph as the Seidel spectrum of any one of the graphs in its switching class. In this context, it can be shown that a two-graph is regular if and only if has two eigenvalues (see [23] for the details).

We suppose therefore that (Ω, Δ) is a regular two-graph with eigenvalues ρ_1 and ρ_2, where $\rho_1 > \rho_2$, say, and that $|\Omega| = v$. Then any ∓ 1 adjacency matrix C of a graph in its switching class satisfies

$$(C - \rho_1 I)(C - \rho_2 I) = 0. \tag{6.1}$$

There are restrictions on ρ_1, ρ_2 that come from the fact that $\text{trace}(C) = 0$, namely that $\rho_1 \rho_2 = 1 - v$ and if $\rho_1 + \rho_2 \neq 0$ then ρ_1 and ρ_2 are odd integers.

If $\rho_1 + \rho_2 = 0$ then $v \equiv 2 \bmod 4$ and $v - 1$ is a sum of squares of two integers. Here $\rho_1 = -\rho_2 = \sqrt{(v-1)}$, which may or may not be integral. Again see [23] for the details.

Corresponding to each $\omega \in \Omega$ there is in the switching class of (Ω, Δ) a graph that has ω as an isolated vertex. Deleting this vertex gives a graph on $v - 1$ vertices, called a *descendant* of (Ω, Δ). Clearly there are at most v non-isomorphic descendants of (Ω, Δ). It is easy to see that every descendant is *strongly regular* with ∓ 1 adjacency matrix B say, satisfying

$$(B - \rho_1 I)(B - \rho_2 I) = -J, \quad B\mathbf{j} = (\rho_1 + \rho_2)\mathbf{j},$$

where, as usual, J and $\mathbf{j}$ are the all-one matrix and all-one vector, respectively. Thus any descendant has the eigenvalue $\rho_0 = \rho_1 + \rho_2$ in addition to ρ_1 and ρ_2 which it inherits from the regular two-graph.

There is another way that a regular two-graph may give rise to strongly regular graphs and that is by switching. It can happen that the switching class of a regular two-graph itself contains *regular* graphs, in which case their ∓ 1 adjacency matrix A must satisfy

$$(A - \rho_1 I)(A - \rho_2 I) = 0, \quad A\mathbf{j} = \rho_1\mathbf{j} \text{ or } A\mathbf{j} = \rho_2\mathbf{j}.$$

Thus A is the adjacency matrix of a strongly regular graph. In this section the only strongly regular graphs we consider are those derived from regular two-graphs by the methods described and for this reason we shall refer to their parameters in the form $(v, \rho_0, \rho_1, \rho_2)$, where $\rho_1 > \rho_2$ are the eigenvalues of the regular two-graph and $\rho_0 = \rho_1$ or $\rho_0 = \rho_2$.

When $n = 36$ we have $\rho_1\rho_2 = -35$, so by considering the complement of the regular two-graph if necessary, we may assume that $\rho_1 = 5$ and $\rho_2 = -7$. Then any graph in the switching class has ∓ 1 adjacency matrix A say, that satisfies $(A - 5I)(A + 7I) = 0$, so that $A + I$ is a symmetric Hadamard matrix of order 36. Moreover, every descendant is a $(35, -2, 5, -7)$ strongly regular graph, and the strongly regular graphs in the switching class have parameters $(36, 5, 5, -7)$ or $(36, -7, 5, -7)$.

It is clear that a regular two-graph is determined by any one of its descendants, for all we have to do is to adjoin an isolated vertex. Thus to determine the regular two-graphs on 36 vertices we need only consider finding strongly regular graphs with parameters $(35, -2, 5, -7)$. The $(0, 1)$ adjacency matrix C of such a graph satisfies

$$C^2 = 9I + 9J, \quad C\mathbf{j} = 18\mathbf{j}. \tag{6.2}$$

To shorten our computer search we have to try to avoid finding two such graphs that are descendants of the same regular two-graph. This is done by gearing our algorithm (as far as we can) to determine just one of the descendants, the one whose standard form is the greatest amongst all the descendants. The idea is to use a backtracking search to construct the $(0, 1)$ adjacency matrix of such a descendant, one entry at a time, until the whole matrix is completed. As in the search for designs described in §4, we have to ensure, as far as economically possible, that we do not descend a path of the search tree that will encounter an isomorphic copy of a graph already found.

Suppose that the adjacency matrix C has been found as far as the rth row:

$$C_r = \begin{bmatrix} A_r & N_r \\ N_r^t & \mathbf{0} \end{bmatrix} \qquad (3 \leq r \leq 35).$$

Then, clearly, if C_r can be completed to a matrix C in standard form, C_r itself must be in standard form. It is important therefore, at least for small values of r, that this be the case. Moreover, we have to ensure as far as economically possible, that C_r will yield a standard form that is maximal. This was achieved by adjoining an isolated vertex to the graph represented by C_r and isolating the vertices 1 to r in turn as long as the standard form of the corresponding descendant was not greater than C_r. If a descendant were found with standard form greater than C_r we backtracked in an attempt to construct a further candidate for the rth row. Lemma 4.1 was useful in this context.

When these tests were incorporated into a procedure written in C many new two-graphs were found relatively quickly, but the exhaustive search is far from being complete. In [2] regular two-graphs on 36 vertices were constructed using Latin squares of order 6 and Steiner triple systems on 15 points. These methods yielded a total of 91 non-isomorphic regular two-graphs. We have taken the 136 new two-graphs that we have found and merged them together in lexicographical order into a list together with the 91 from [2], giving a total of 227 regular two-graphs [34]. As a result of the (so far incomplete) computer search outlined above, we can say that all the regular two-graphs that lie lexicographically between numbers 1 and 225 have been found.

As pointed out earlier, these regular two-graphs may have strongly regular graphs in their switching classes. Here the graphs are actually $(36, 15, 6)$ and $(36, 21, 12)$ graphs. We have determined completely the numbers of these coming from the 227 regular two-graphs, finding 32548 and 180 respectively. In the first case, amongst the numbers found there are 25634 that have trivial automorphism group, and hence by the result of [8] referred to in §3, there are at least 25634 non-isomorphic 2-$(36, 15, 6)$ designs (with trivial automorphism group). Moreover, the regular two-graphs give rise to 3854 non-isomorphic descendants, of which 2240 have trivial auto-morphism group. Of course these descendants are also 2-(35, 18, 9) designs, perhaps not all pairwise non-isomorphic. However, as pointed out earlier, if A is the $\mp$ adjacency matrix of any representative of a regular two-graph on 36 vertices, then $A + I$ is a symmetric Hadamard matrix with constant diagonal. We investigated just how many of the 227 regular two-graphs that we found were pairwise non-isomorphic as Hadamard matrices, i.e., under the operations of row and column permutations and the multipli-cation of rows and/or columns by -1. We discovered in total there to be 180 non-isomorphic Hadamard matrices and, as in §5, we analysed these for Hadamard designs (in this case 2-(35, 17, 8) designs, the complements of 2-(35, 18, 9) designs) and found 108131. This explains the corresponding entry in the table featured in the introduction.

It was as a result of an observation initially made by Bussemaker [4] on the output of the computer program and contributions from J. J. Seidel [25], that the author was able to find a new classification of 100 of the above 227 regular two-graphs.

Let P be a symmetric conference matrix of order 10, so that $P^2 = 9I$, and put

$$A = \begin{bmatrix} I+P & I-P \\ I-P & I+P \end{bmatrix} \text{ giving } A^2 = \begin{bmatrix} 20I & -16I \\ -16I & 20I \end{bmatrix}.$$

Assume that A can be embedded as a principal sub-matrix of order 20 of a symmetric Hadamard matrix H of order 36 with diagonal I. Thus

$$H = \begin{bmatrix} A & N \\ N^t & B \end{bmatrix} \text{ satisfies } H^2 = 36I.$$

Write N in the form $\begin{bmatrix} N_1 \\ N_2 \end{bmatrix}$, where N_i is of size 10×16 ($i = 1, 2$). Then from the relation $NN^t = 16 \begin{bmatrix} I & I \\ I & I \end{bmatrix}$, it follows that $N_i N_j^t = 16I$ ($i, j = 1, 2$). Clearly $(N_1 - N_2)(N_1 - N_2)^t = 0$, so that $N_1 = N_2$. Write $M = N_1 = N_2$.

Theorem 6.1 *There exist diagonal ± 1 matrices D_1 and D_2 such that*

$$M = D_1 [\, C \ \mathbf{j}\,] D_2,$$

where C is the ± 1 incidence matrix of a 2-(10, 4, 2) design.

Since any 2-(10, 4, 2) design is *quasi-symmetric* with intersection numbers 1 and 2 the converse is easily established:

Theorem 6.2 *If C is the ± 1 incidence matrix of a 2-(10, 4, 2) design and D_1, D_2 are diagonal ± 1 matrices of order 10, 16 respectively, and if $M = D_1 [C \quad \mathbf{j}] D_2$, and $B = 6I - \frac{1}{2} M^t M$, then B is a symmetric ± 1 matrix,*

$$H = \begin{bmatrix} I+P & I-P & M \\ I-P & I+P & M \\ M^t & M^t & B \end{bmatrix}$$

is a symmetric Hadamard matrix of order 36, and $H - I$ is a regular two-graph on 36 vertices.

Gronau [7] has proved that there are exactly 3 pairwise non-isomorphic 2-(10, 4, 2) designs. We have used these three designs and have identified the 100 non-isomorphic regular two-graphs that can be constructed from them by the method just described. Fuller details are given in [34].

Another theorem that was discovered by examining the output of a computer program is the one that follows, where we construct new regular two-graphs from old. Here the regular two-graphs involved are not confined to having 36 vertices.

Let Γ be a regular two-graph on v vertices and let C be the ∓ 1 adjacency matrix of a graph in its switching class. Then, as before,

$$(C - \rho_1 I)(C - \rho_2 I) = 0, \quad \rho_1 > \rho_2.$$

Let $K = C - \frac{1}{2}(\rho_1 + \rho_2) I$, and $m = (\rho_1 - \rho_2)/2$. Then K is symmetric, has constant diagonal and has eigenvalues $\pm m$. Suppose now that A is a principal sub-matrix of K (of any size) and that K is partitioned according to

$$K = \begin{bmatrix} A & N \\ N^t & B \end{bmatrix},$$

so that

$$A^2 + NN^t = m^2 I, \quad AN + NB = 0, \quad N^t N + B^2 = m^2 I.$$

Theorem 6.3 *Let Q be a generalised permutation matrix that commutes with A^2 and suppose that for every eigenvalue α of A, $(\alpha \neq \pm m)$, $-\alpha$ is not an eigenvalue of A. Then the matrix K_Q defined by*

$$K_Q = \begin{bmatrix} A & QN \\ (QN)^t & B \end{bmatrix}$$

is symmetric and satisfies $K_Q^2 = m^2 I$. Moreover, $K_Q + \left(\dfrac{\rho_1 + \rho_2}{2}\right) I$ is in the switching class of a regular two-graph on v vertices.

Details of the proof can be found in [34]

§7 A family of symmetric 2-designs

Until fairly recently it was an unsolved problem as to whether or not there exists a symmetric 2-$(160, 54, 18)$ design. Its existence is not ruled out by the Bruck-Ryser-Chowla Theorem but its parameters seemed to be too large to enable an attempt at direct construction. However, we shall see that this is not the case and we shall indicate how an attempt to construct this design [31] led to the establishment of a new family [32], [21].

Let us assume that there exists a 2-$(160, 54, 18)$ design having a fixed-point-free automorphism σ of order 4. Then there are 40 orbits of points and 40 orbits of blocks, all of length 4. If we proceed as in §2 to construct an orbit matrix Q corresponding to σ, we find that Q must satisfy the conditions

$$\sum_{\ell=1}^{40} Q_{i\ell} = 54, \qquad \sum_{\ell=1}^{40} Q_{i\ell}Q_{j\ell} = 72,$$

$1 \leq i, j \leq 40$, $i \neq j$. If one makes the further assumption that the entries of Q are either 0 or 2, the clearly $Q/2$ is the incidence matrix of a symmetric 2-$(40, 27, 18)$ design. Thus we are led to the question: does there exist a 2-$(40, 27, 18)$ design in whose $(0, 1)$ incidence matrix the ones can be replaced by 4 circulant (cyclic) matrices with row (and column sums) 2, and whose zeros can be replaced by the zero 4×4 matrix, in such a way that the resulting matrix of size 160 is the incidence matrix of a 2-$(160, 54, 18)$ design?

Since the author had several hundred of these 2-$(40, 27, 18)$ designs as a result of [27], there was a possibility that one of them might be a candidate. However, when the computer programs used in the earlier examples were adapted to this case, it was clear that unless one was extremely lucky and fell upon a solution, the search was hopeless. One reason for this was that the programs were geared to an *exhaustive search* and here we were realistically only looking for one solution. The progress made by the programs suggested that even rewriting them to cater for this new situation would prove a reward-less task. Thus, in this respect, the computer was not a useful tool.

There was a feeling that if any 2-$(40, 27, 18)$ design was going to provide a solution then it would be the complementary design of the 2-$(40, 13, 4)$ design defined by the hyper-planes in $PG(2, 3)$, and indeed this turned out to be the case. The key fact turned out to be the realisation that the cyclic form of the incidence matrix of this design could be used to determine three *negacyclic* matrices C_1, C_2, C_3 of size 40 with entries $0, \pm 1$, such that

$$C_i C_i^t = 9I, \quad (1 \leq i \leq 3) \text{ and } C_i * C_j = 0, \quad (1 \leq i < j \leq 3),$$

with the further condition that the matrix C defined by $C = C_1 + C_2 + C_3$ satisfies

$$CC^t = 27I.$$

If the -1 entries of C are replaced by 1 the resulting matrix is the incidence matrix of the underlying 2-$(40, 27, 18)$ design. Here, the notation $C_i * C_j$ denotes the Hadamard product obtained by multiplying corresponding entries.

Further ingredients for the construction of the design were the three 4×4 matrices H_1, H_2, H_3 that correspond to the three subgroups of GF(4) of order 2,

$$
H_1 = \begin{bmatrix} 1 & 1 & 0 & 0 \\ 1 & 1 & 0 & 0 \\ 0 & 0 & 1 & 1 \\ 0 & 0 & 1 & 1 \end{bmatrix}, \quad
H_2 = \begin{bmatrix} 1 & 0 & 1 & 0 \\ 0 & 1 & 0 & 1 \\ 1 & 0 & 1 & 0 \\ 0 & 1 & 0 & 1 \end{bmatrix}, \quad
H_3 = \begin{bmatrix} 1 & 0 & 0 & 1 \\ 0 & 1 & 1 & 0 \\ 0 & 1 & 1 & 0 \\ 1 & 0 & 0 & 1 \end{bmatrix}.
$$

It is easy to see that

$$
\begin{aligned}
H_1 + H_2 + H + 3 &= 2I + J, \\
H_i H_j &= J \ (i \neq j), \\
H_i^2 &= 2H_i, \\
H_i(J - H_j) &= J \ (i \neq j), \\
H_i(J - H_i) &= 2(J - H_i).
\end{aligned}
$$

Although these matrices are not circulants as we originally requested, the above relationships ensure that they can be inserted in the matrix C to yield a 2-$(160, 54, 18)$ design. The rule is that whenever an entry $+1$ in C is encountered that arises from an entry in C_i it is replaced by the matrix H_i, an entry -1 arising from an entry in C_i is replaced by $J - H_i$ and all zeros are replaced by the zero 4×4 matrix. The design obtained has full automorphism group the Klein four group.

Many non-isomorphic designs with the same parameters 2-$(160, 54, 18)$ can be constructed using minor modifications to the above method. It is possible, for example, to replace the matrices H_1, H_2, H_3 by circulants to get a design with full automorphism group the cyclic group of order 4.

By using the concept of a relative difference set the above result has been extended to produce an infinite family of new symmetric 2-designs [32] and this family itself was later extended by Pott and Jungnickel in [21].

References

[1] W.G.Bridges, M.Hall Jr. and J.L.Hayden, Codes and Designs, *J. Combin. Theory Ser. A* **31** (1981), 155–174.

[2] F.C.Bussemaker, R.A.Mathon and J.J.Seidel, Tables of two-graphs, *Combinatorics and Graph Theory*, Lecture Notes in Mathematics, (S.B.Rao, ed.), **885** Springer, 1981, 70–112; *Report 79-WSK-05*, Techn. Univ. Eindhoven, 1979.

[3] F.C.Bussemaker, W.H.Haemers, J.J.Seidel and E.Spence, On (v, k, λ) graphs and designs with trivial automorphism group, *J. Combin. Theory Ser. A* **50** (1989), 33-46.

[4] F.C.Bussemaker, Private communication, 1990.

[5] V.Čepulić, On symmetric block designs $(40, 13, 4)$ with automorphisms of order 5, *Discrete Math.*, **128** (1994) 45–60,

[6] R.H.F.Denniston, Enumeration of symmetric designs $(25, 9, 3)$, *Ann. Discrete Math.* **15** (1982), 111–127.

[7] H.-D.O.F. Gronau, The 2-(10,4,2) Designs, *Rostock Math. Kolloq.* **16** (1981), 5–10.

[8] W.H.Haemers, Dual Seidel switching, in Papers dedicated to J.J.Seidel *(P.J.de Doelder, J.de Graaf and J.H.van Lint, Eds.)*, pp. 183–190, Technical University Eindhoven, 1984.

[9] W.H.Haemers and E.Spence, On (v, k, λ) graphs and designs without involution, *COMBINATORICS' 88, Proceedings of the International Conference on Incidence Geometries and Combinatorial Structures* (Ravello, Italy, 1988), **2** (1991), 437-447

[10] M.Hall Jr., Combinatorial Theory (1986), John Wiley & Sons.

[11] N.Ito, J.S.Leon and J.Q.Longyear, Classification of 3-(24, 12, 5) designs and 24-dimensional Hadamard matrices, *J. Combin. Theory Ser. A* **27** (1979), 289–306.

[12] H.Kimura, Hadamard matrices of order 28 with automorphism groups of order 2, *J. Combin. Theory Ser. A* **43** (1986), 98–102.

[13] H.Kimura, New Hadamard Matrix of order 24, *Graphs and Combin.* **5** (1989), 139–146.

[14] H.Kimura, Characterisation of Hadamard matrices of order 28 with Hall sets, *Discrete Math.*, **128** (1994) 257–268.

[15] H.Kimura, Classification of Hadamard matrices of order 28, *Discrete Math.*, **133** (1994) 171–180.

[16] C.W.H.Lam, L.Tiel and S.Swiercz, The non-existence of finite projective planes of order 10, *Canad. J. Math* **41** (1989), 1117–1123.

[17] E.S.Lander, Symmetric designs: an algebraic approach, London Mathematical Society Lecture Note Series 74 (1983), Cambridge University Press.

[18] R.Mathon, Symmetric (31, 10, 3)-designs with non-trivial automorphism group, *Ars Combinatoria* **25** (1988), 171–183.

[19] R.Mathon and A.Rosa, Tables of parameters of BIBD's with $r \leq 41$ including existence, enumeration and resolvability results: an update, *Ars Combinatoria* **30** (1990), 65–96.

[20] R.Mathon and E.Spence, On $(45, 12, 3)$ designs, to appear.

[21] A.Pott and D.Jungnickel, A new family of symmetric block designs, *Designs Codes and Cryptography*

[22] A.Rudvalis, (v, k, λ) graphs and polarities of (v, k, λ) designs, *Math. Zeitschr.* **20** (1971), 224–230.

[23] J.J.Seidel, A survey of two-graphs, *Coll. Intern. Teorie Combin.*, Atti dei convegni Lincei 17, Roma, (1976), 481–511.

[24] J.J.Seidel, D.E.Taylor, Two-graphs, a second survey, Algebr. Methods in Graph Theory (L.Lovász, Vera T.Sós, eds.) *Coll. Math. Soc. J.Bolyai*, **25** (1981), 698–711.

[25] J.J.Seidel, Private communication, 1990.

[26] J.J.Seidel, More about two-graphs, *Fourth Czechoslovakian Symposium on Combinatorics, Graphs and Complexity*, (J.Nešetřil and M.Feidler eds.), (1992), 297–308.

[27] E.Spence, (40,13,4) designs derived from strongly regular graphs, *Advances in Finite Geometries and Designs*, (J.W.Hirschfeld, D.R.Hughes, J.A.Thas eds.) Oxford University Press 1991, 359–368.

[28] E.Spence, Symmetric $(31, 10, 3)$ designs with a non-trivial automorphism of odd order, *Journ. Comb. Math. and Comb. Designs* **10**, (1991), 51-64.

[29] E.Spence, A complete classification of symmetric $(31, 10, 3)$ designs, *Designs, Codes and Cryptography*, **2** (1992), 127–136.

[30] E.Spence, Symmetric $(41, 16, 6)$ designs with a non-trivial automorphism of odd order, *J. Combin. Designs* **1** (1993) 193–211.

[31] E.Spence, V.D.Tonchev and T.van Trung, A symmetric $(164, 54, 18)$ design, *J. Combin. Designs* **1** (1993) 65–68.

[32] E.Spence, A new family of symmetric 2-(v, k, λ) block designs, *Europ. J. Combinatorics* **14** (1993) 131–136.

[33] E.Spence, Classification of Hadamard matrices of orders 24 and 28, to appear in *Discrete Math*.

[34] E.Spence, Regular 2-graphs on 36 vertices, to appear.

[35] E.Spence, Unpublished computer result (1994).

[36] V.D.Tonchev, Hadamard matrices of order 28 with an automorphism of order 13, *J. Combin. Theory Ser. A* **35** (1983), 43–57.

[37] V.D.Tonchev, Hadamard matrices of order 28 with an automorphism of order 7, *J. Combin. Theory Ser. A* **40** (1985), 62–81.

[38] V.D.Tonchev, Symmetric 2-$(31, 10, 3)$ designs with an automorphism of order seven, "Combinatorial Design Theory", North-Holland, Amsterdam-New York, (1987), 461–464.

Modern Probabilistic Methods in Combinatorics

Joel Spencer

The *probabilistic method* is a means to prove the existence of configurations by showing that an appropriately defined random configuration has a positive probability of having the desired property. The method is approaching its golden anniversary, its beginning generally considered a three-page paper by Paul Erdős [6] in 1947. Closely aligned is the study of *random graphs*, more generally random configurations, in which problems about probabilities concerning random graphs are considered for their own sake. This topic began in 1961 with the monumental study of Paul Erdős and Alfred Rényi [9], "On the evolution of random graphs". For many years the uses of probability in these twin topics was surprisingly elementary, linearity of expectation, variance and the Chernoff bounds could take a fledgling researcher a long long way. Recent years have seen more sophisticated uses of probability and our emphasis here will be on the newer probabilistic methodologies and how they are applied to these topics. We give our recent book [2] and the book of Bollobás [3] on Random Graphs as a general references.

1 Exponential Haystacks

1.1 Janson Inequalities

Let $A_1, \ldots, A_m$ be events in a probability space. Set

$$M = \prod_{i=1}^{m} \Pr[\overline{A_i}]$$

The Janson Inequality allows us, sometimes, to estimate $\Pr[\wedge \overline{A_i}]$ by M, the probability if the A_i were mutually independent. The original proof by Svante Janson is in [13]. See [5] for a more "elementary" proof and [2] for general discussion. We let G be a dependency graph for the events – i.e., the vertices are the indices $i \in [m]$ and each A_i is mutually independent of all A_j with j not adjacent to i in G. (This notion was first used with the Lovász Local

Lemma. While the dependency graph is not uniquely defined there is usually a clear candidate.) We write $i \sim j$ when i, j are unequal and adjacent in G. We set

$$\Delta = \sum_{i \sim j} \Pr[A_i \wedge A_j]$$

We make the following *correlation assumptions*:
• for all i, S with $i \notin S$

$$\Pr[A_i | \wedge_{j \in S} \overline{A}_j] \leq \Pr[A_i]$$

• for all i, k, S with $i \neq k$ and $i, k \notin S$

$$\Pr[A_i \wedge A_k | \wedge_{j \in S} \overline{A}_j] \leq \Pr[A_i \wedge A_k]$$

Finally, let ϵ be such that $\Pr[A_i] \leq \epsilon$ for all i.
The Janson Inequality: Under the above assumptions

$$M \leq \Pr[\wedge \overline{A}_i] \leq M e^{\frac{\Delta}{2(1-\epsilon)}} \tag{1}$$

We set

$$\mu = \sum \Pr[A_i],$$

the expected number of A_i that occur. As $1 - x \leq e^{-x}$ for all $x \geq 0$ we may bound $M \leq e^{-\mu}$ and then rewrite the upper bound in the somewhat weaker but quite convenient form

$$\Pr[\wedge \overline{A}_i] \leq e^{-\mu + \frac{\Delta}{2(1-\epsilon)}}$$

In most applications $\epsilon = o(1)$ and the pesky factor of $1 - \epsilon$ is no real trouble. Indeed just assuming all $\Pr[A_i] \leq \frac{1}{2}$ is plenty for all cases we know of. In many cases we also have $\Delta = o(1)$. Then the Janson Inequality gives an asymptotic formula for $\Pr[\wedge \overline{A}_i]$. When $\Delta \gg \mu$, as also occurs in some important cases, the above gives an upper bound for $\Pr[\wedge \overline{A}_i]$ which is bigger than one. In those cases we sometimes can use the following:
The Extended Janson Inequality: Under the assumptions of the Janson Inequality and the additional assumption that $\Delta \geq \mu(1 - \epsilon)$

$$\Pr[\wedge \overline{A}_i] \leq e^{-\frac{\mu^2(1-\epsilon)}{2\Delta}} \tag{2}$$

In our application the underlying probability space will be the random graph $G(n, p)$. The events A_α will all be of the form that $G(n, p)$ contains a particular set of edges E_α. The correlation assumptions are then an example of far more general result called the *FKG inequalities*. We have a natural dependency graph by making A_α, A_β adjacent exactly when $E_\alpha \cap E_\beta \neq \emptyset$.

Let us parametrize $p = c/n$ and consider the property , call it TF, that G is triangle free. Let A_{ijk} be the event that $\{i, j, k\}$ is a triangle in G. Then

$$TF = \wedge \overline{A}_{ijk},$$

the conjunction over all triples $\{i, j, k\}$. We calculate

$$M = \left(1 - p^3\right)^{\binom{n}{3}} \sim e^{-\mu}$$

with $\mu = \binom{n}{3}p^3 \sim c^3/6$. We bound Δ by noting that we only need consider terms of the form $A_{ijk} \wedge A_{ijl}$ as otherwise the edge sets do not overlap. There are $O(n^4)$ choices of such i, j, k, l. For each the event $A_{ijk} \wedge A_{ijl}$ is that a certain five edges (ij, ik, jk, il, jl) belong to $G(n, p)$, which occurs with probability p^5. Hence

$$\Delta = \sum \Pr[A_{ijk} \wedge A_{ijl}] = O(n^4 p^5)$$

With $p = c/n$ we have $\epsilon = O(n^{-3}) = o(1)$ and $\Delta = o(1)$ so that the Janson Inequality gives an *asyptotic formula*

$$\Pr[TF] \sim M \sim e^{-\frac{c^3}{6}}$$

This much could already be done in the original work of Erdős and Rényi by calculation of moments. But the Janson Inequalities allow us to proceed beyond $p = \Theta(1/n)$. The calculation $\Delta = o(1)$ had plenty of room. For any $p = o(n^{-4/5})$ we have $\Delta = o(1)$ and therefore an asymptotic formula $\Pr[TF] \sim M$. For example, if $p = \Theta\left((\ln n)^{1/3}/n\right)$ this yields that $G(n, p)$ has polynomially small probability of being trianglefree. Once p reaches $n^{-4/5}$ the value Δ becomes large and we no longer have an asymptotic formula. But as long as $p = o(n^{-1/2})$ we have $\Delta = O(n^4 p^5) = o(n^3 p^3) = o(\mu)$ and so we get the *logarithmically asymptotic* formula

$$\Pr[TF] = e^{-\mu(1+o(1))} = e^{-\frac{n^3 p^3}{6}(1+o(1))}$$

Once p reaches $n^{-1/2}$ we lose this formula. But now the Extended Janson Inequality comes into play. We have $\mu = \Theta(n^3 p^3)$ and $\Delta = \Theta(n^4 p^5)$ so for $p \gg n^{-1/2}$

$$\Pr[TF] < e^{-\Omega(\mu^2/\Delta)} = e^{-\Omega(n^2 p)}$$

The Extended Janson Inequality gives, in general, only an upper bound. In this case, however, we note that $\Pr[TF]$ is at least the probability that $G(n, p)$ has no edges whatsoever and so, for $n^{-1/2} \ll p \ll 1$

$$\Pr[TF] > (1 - p)^{\binom{n}{2}} = e^{-\Omega(n^2 p)}$$

With a bit more care, in fact, one can estimate $\Pr[TF]$ up to a constant in the logarithm for all p. These methods do not work just for trianglefreeness. In a remarkable paper Andrzej Rucinski, Tomasz Luczak and Svante Janson [13] have examined the probability that $G(n,p)$ does not contain a copy of H, where H is any particular fixed graph, and they estimate this probability, up to a constant in the logarithm, for the entire range of p. Their paper was the first and is still one of the most exciting applications of the Janson Inequality.

1.2 Martingale Inequalities

Martingales have a long history in probability theory but their usefulness in our context is quite new. We refer to Colin McDiarmid's excellent survey [16] at this meeting for a more detailed examination. For our purposes we consider a martingale to be a sequence $X_0, \ldots, X_m$ of random variables (on a common space) so that for any $0 \le i < m$ and value a $E[X_{i+1}|X_i = a] = a$. We further assume $X_0 = \mu$, a constant. Then $\mu = E[X_i]$ for all i.

Azuma's Inequality: Let $\mu = X_0, X_1, \ldots, X_m = X$ be a martingale in which $|X_{i+1} - X_i| \le 1$. Then for any $a > 0$

$$\Pr[X > \mu + a] < \exp(-a^2/2m) \qquad (3)$$

In application we use an isoperimetric version. Let $\Omega = \prod_{i=1}^m \Omega_i$ be a product probability space and X a random variable on it. Call X *Lipschitz* if whenever $\omega, \omega' \in \Omega$ differ on only one coordinate $|X(\omega) - X(\omega')| \le 1$. Set $\mu = E[X]$.

Azuma's Perimetric Inequality: $\Pr[X \ge \mu + a] < e^{-2a^2/m}$.

The connection is via the Doob Martingale, $X_i(\omega)$ being the conditional expectation of X given the first i coordinates of ω. The same inequality holds for $\Pr[X \le \mu - a]$. Direct application of Azuma's Inequality gives only an $e^{-a^2/2m}$ bound, see [16] for this improved result.

E. Shamir and this author [19] applied this result to the chromatic number $\chi(G)$ of the random graph $G \sim G(n,p)$. (Again [2] gives a general discussion.) Let Ω be the probability space, whose vertices, i.e., graphs, may be thought of as Boolean arrays of length $\binom{n}{2}$. Let $X : \Omega \to Z$ be chromatic number. For $2 \le i \le n$ let Ω_i be the restriction of the graph to the pairs $\{j, i\}, 1 \le j < i$. We may think of Ω_i as $i - 1$ values of the full Boolean array or as the "information" about vertex i looking to the "left". Now X is Lipschitz since we can make any change to the edges involving vertex i and it can only increase X by at most one since we can always give i a new color. This yields a strong concentration result:

$$\Pr[|\chi(G) - \mu| \ge \lambda(n-1)^{1/2}] < 2e^{-2\lambda^2} \qquad (4)$$

so that, roughly, the chromatic number is concentrated within $n^{1/2}$ of its expectation. An oddity of this method is that it does not by itself give the value of the expectation, it only deduces that the random variable is tightly concentrated around its expectation.

We can generalize 4 considerably. We call a graph function X vertex Lipschitz if changing the edges at one vertex can only change $X(G)$ by at most one. Then 4 holds with χ replaced by any vertex Lipschitz X. Further we can alter the probability measure (holding to the set of graphs on $\{1, \ldots, n\}$ as our objects) as long as the component parts Ω_i are mutually independent. For example, let H be a fixed (not necessarily random) graph on $\{1, \ldots, n\}$ and let H_p denote the random subgraph given by selecting edges from H with independent probability p and selecting no edges outside of H. Then the distribution $\chi(H_p)$ satisfies the concentration 4. Somewhat more generally suppose for each $1 \leq i < j \leq n$ there is a $p_{ij} \in [0,1]$ and consider the random graph G with i,j adjacent with probability p_{ij}, the adjacencies mutually independent events. Again, for any choice of p_{ij} and any vertex Lipschitz X the random variable $X(G)$ satisfies the concentration 4.

We call graph function X edge Lipschitz if changing any single edge (from in to out or out to in) can change $X(G)$ by at most one. Set $m = \binom{n}{2}$. We can decompose $G \sim G(n,p)$ as the product of m Binary choices so that Azuma's Inequality gives

$$\Pr[|X(G) - \mu| \geq \lambda m^{1/2}] < 2e^{-2\lambda^2} \tag{5}$$

where $\mu = E[X]$. Bollobás [4] used this to give a remarkable bound on the clique number $\omega(G)$. Fix $p = \frac{1}{2}$ for definiteness. Set Y_k equal the number of k-cliques and

$$f(k) = E[Y_k] = \binom{n}{k} 2^{-\binom{k}{2}}$$

Elementary analysis shows that $f(k_0) > 1 > f(k_0 + 1)$ for some $k_0 \sim 2 \log_2 n$ and, for $k \sim k_0$, $f(k+1)/f(k) = n^{-1+o(1)}$. As $\Pr[\omega(G) \geq k] \leq E[Y_k]$ almost surely $\omega(G) < k_0 + 2$. Now set $k^- = k_0 - 3$ so that $f(k^-) > n^{3-o(1)}$. Bollobás showed

$$\Pr[\omega(G) < k^-] < 2^{-cn^2 \ln^{-8} n} \tag{6}$$

This is "near" best possible in that with probability 2^{-cn^2} the graph has no edges whatsoever. To prove this set X equal to the maximal size of a family of edge-disjoint cliques of size k^-. Note $X = 0$ if and only if $\omega(G) < k^-$ and that X is edge Lipschitz. From less modern (though nontrivial) probabilistic methods one can show $\mu = E[X] > cn^2 \ln^{-4} n$. Now 6 follows from 5 by setting $\lambda = \mu m^{-1/2}$. Its interesting to note that the same result (with a different power of $\ln n$ in the exponent) can be derived directly from the Extended Janson Inequality. From this Bollobás showed that the chromatic number $\chi(G)$ was almost surely $\sim 2 \log_2 n$.

We conclude with a variant of Azuma's Inequality used in the work J.H. Kim discussed in §2.2. Let I_i, $1 \le i \le m$, be mutually independent identically distributed indicator random variables with $E[I_i] = p$. (E.g., I_i is the indicator for the i-th edge in $G(n, p)$.) Let X be a function of the I_i (e.g., a graph function) such that changing I_i can change X by at most c_i. Set $\sigma = [p \sum c_i^2]^{1/2}$. (If $X = \sum c_i I_i$ then $Var[X] < \sigma^2$ and σ is like a standard deviation.) Then

$$\lambda \max(c_i) \le 2\sigma^2 \ln 2 \Rightarrow \Pr[|X - E[X]| \ge \lambda] < 2e^{-\lambda^2/4\sigma^2} \qquad (7)$$

For $p = o(1)$ this is much tighter that the basic Azuma bound and the use of the c_i allows a clear sense of the weighting of influences of different potential edges.

2 Dynamic Algorithms

2.1 Asymptotic Packing

For $2 \le l < k < n$ let $m(l, k, n)$ be the maximal size of a family of P of k-element subsets of an n-set Ω such that every l points lie in *at most* one $A \in P$. Such P are naturally called packings. Our concern will be for l, k fixed, $n \to \infty$. Elementary counting gives

$$m(l, k, n) \le \frac{\binom{n}{l}}{\binom{k}{l}} \qquad (8)$$

with equality if and only if there is a design with every l points lying in a unique $A \in P$. For $l = 2, k = 3$ these are the famous Steiner Triple Systems and for $l = 2$ and any fixed k now classic results of R. Wilson give asymptotic necesary and sufficient conditions for the existence of such designs. The situation for $l > 2$ is much less well understood. In 1961 Paul Erdős and Haim Hanani [8] asked whether 8 holds asymptotically – i.e.:

$$\lim_{n \to \infty} m(l, k, n) \frac{\binom{k}{l}}{\binom{n}{l}} = 1 \qquad (9)$$

This conjecture was proven by V. Rödl [18] in 1985 by a technique often called the Rödl nibble.

Recent years have seen a reevaluation of Rödl's Theorem from the viewpoint of random dynamic algorithms. Take all $\binom{n}{k}$ k-sets and order them randomly. Now create a packing P dynamically, beginning with $P = \emptyset$. We consider the k-sets E in order. We add E to P if possible. More precisely,

E is added to P if and only if there is no F already in P overlapping E in at least l points. This certainly will create a packing P but the real result is that P will have expected size as desired.

We turn this into a continuous time dynamic process as follows. To each of the $\binom{n}{k}$ k-sets E we assign a birthtime x_E. The x_E are chosen independently, each a uniformly chosen real number in $[0, \binom{n-l}{k-l}]$. (This choice of interval length will make calculations convenient shortly.) Time starts at zero with $P = \emptyset$. When E is born it is added to P if possible, as before. Of course, the E are considered in random order so that the *final* value of P has the same distribution as before. We consider P_c, the value of P at time c, where c is a fixed real. We say an l-set e is covered at time c if $e \subset E$ for some $E \in P_c$. We now want the probability e is so covered.

We define a continuous time branching process that mirrors the fate of e above. Begin at time c with a single "Eve". Time goes continuously backwards to zero. Eve gives birth with a unit density Poisson process – in infintesmal time dt she has probability dt of giving birth. All births are to precisely Q children where $Q = \binom{k}{l} - 1$, the children in the same birth are called wombmates. (Littermates is the biologist's term for animals but English lacks a word for humans except when $Q = 2$. Note "siblings" is quite different.) The children are born mature and have births by the same random process as do their children and so forth. A rooted tree T is thus generated and it can be shown that with probability one T is finite. We call vertices of T surviving or dying as follows. All childless vertices are surviving. A vertex is dying if and only if it has a (at least one) birth all of whose wombmates are surviving. Working up from the leaves of the tree every vertex of T is so designated. Let $g(c)$ denote the probability that the root Eve survives.

We claim that the limit (as $n \to \infty$) of the probability e is not covered is $g(c)$. To see (informally) the mirror fix an l-set e and start at time c, time going backwards. Identify e with Eve. When a k-set $A \supset e$ is born consider this a birth of the Q l-sets $f \subset A$, $f \neq e$. There are $\binom{n-l}{k-l}$ potential births so in infintesmal time dt there is probability dt of having such a birth. (The total number of births is given by a Binomial distribution and a central aspect of the asymptotics is estimation of the Binomial by the Poisson.) Once f has been born the birth of a k-set $B \supset f$ is considered as a birth of the Q l-sets $g \subset B$, $g \neq f$. A tree T is thus generated. (Actually, it *may* happen that a k-set A is born which contains two (or more) l-sets f in which case our analogy fails. This, however, can be shown to occur with probability $o(1)$.) T determines if e is covered. If T consists only of root e then no $A \supset e$ were born so e is not covered. We show by induction on the size of T that e is not covered if and only if it survives in T as defined above. If e survives then for every $A \supset e$ there is an $f \subset A$, $f \neq e$ that does not survive. The rooted subtree at f is the tree generated starting at f at time x_A. By induction f did not survive so there was a $B \supset f$ with $x_B < x_A$ that was added to P. Then at time x_A A was not added to P. This holds for all A so e is not covered at time c. Inversely if e does not survive then there is an $A \supset e$ so that all $f \subset A$, $f \neq e$, do survive. By induction at time x_A no such f has been covered. Either e is already covered or A is now placed in P, so e is covered by or at time x_A, either way e is covered by time c.

Now we can focus attention on $g(c)$, a totally continuous problem for which the pesky n has disappeared. We consider this a function of c and compare $g(c)$ with $g(c + dc)$ for infinitesmal dc. The difference in Eve's survival chances are if she had no births up to time c for which all children survived but then has a birth in the time interval $[c, c + dc)$ for which all children survive. Thus $g(c) - g(c + dc)$ is roughly $g(c)(dc)g(c)^Q$, reflecting the three factors. This can be made precise (we've skipped the necessary first step, showing that $g(c)$ is continuous) and g can be shown to satisfy the differential equation

$$g'(c) = -g(c)^{Q+1} \tag{10}$$

Together with the initial condition $g(0) = 1$ this has the unique solution

$$g(c) = (1 + Qc)^{-1/Q} \tag{11}$$

As $\lim_{c \to \infty} g(c) = 0$ this gives a proof of Rödl's Theorem.

One can put these results into a more general (and perhaps more natural) context. Let H be a $Q + 1$-uniform hypergraph on v vertices. Suppose H is nearly regular in the sense that $\deg(e) \sim D$ for every vertex e where $D \to \infty$. Define the *codegree* of e, f to be the number of edges containing them both and assume that all codegrees are $o(D)$. (Formally, we may consider an infinite sequence of such hypergraphs, Q fixed, with asymptotics defined as the structures become bigger.) N. Pippenger (as given in [17]) has shown that under these circumstances there exists a packing P of $\sim v/(Q + 1)$ disjoint edges. To translate the Erdős-Hanani situation into this context create a hypergraph H_n whose vertices are the l-element subsets of $\Omega = \{1, \ldots, n\}$ and whose edges are $\{e \subset E : |e| = l\}$ for each k-set $E \subset \Omega$. Then $Q+1 = \binom{k}{l}$, H is regular with $D = \binom{n-l}{k-l}$ and the codegrees are all $O(n^{k-l-1}) = o(D)$. The proof of Pippenger's generalization via continuous time branching processes is essentially as before. Now we give each edge E a birthdate unifromly distributed in $[0, D]$. Again given a vertex e and time c we generate a tree T to determine if e is covered. Again the analogy may fail and the condition on the codegrees turns out to be precisely what is neded to show that this occurs with probabiity $o(1)$.

Now, sticking with the hypergraph format, we describe another means toward the same end. Again we have a $Q + 1$-uniform hypergraph with all

$\deg(v) \sim D$ and all codegrees $o(D)$, to each edge E we assign a uniformly distributed birthdate $x_E \in [0, D]$ and let $P = P_t$ be the packing at time t. Let $L = L_t$ be the complement of $\bigcup P$. Let $S = S_t$ (for surviving) be the restriction of H to L_t. Let $\deg_t(v)$ denote the degree of v in S_t, defined for $v \in L_t$. The idea now is to find a function $f(t)$ so that almost surely most $\deg_t(v) \sim f(t)D$.

Suppose there is such a function $f(t)$. Consider the evolution of S from t to $t + dt$ with respect to a vertex $v \in L_t$. Most $w \in L_t$ lie in $\sim f(t)D$ edges of S_t and each edge is born with probability dt/D (and if born it is added to P) so with probability $\sim f(t)dt$ w is removed from L. Consider an edge $E \in S_t$ containing v. Conditioning on v itself remaining in L there is probability $\sim Qf(t)dt$ that $E \notin S_{t+dt}$ as any of the other vertices could be removed. Thus the degree of v will drop by an expected amount $(f(t)D)(Qf(t)dt)$, giving $\deg_{t+dt}(v) \sim D(f(t) - Qf^2(t)dt)$. This yields the differential equation

$$f'(t) = -Qf^2(t) \tag{12}$$

for f which, given the initial condition $f(0) = 1$, has the unique solution

$$f(t) = (1 + Qt)^{-1} \tag{13}$$

Now let $g(t)$ be the probability that $v \in L_t$. Given $v \in L_t$ it has probability $\sim f(t)dt$ of being in an edge now placed in P so that $g(t+dt) \sim g(t)(1 - f(t)dt)$. Letting $h(t) = \ln g(t)$ we have $h(t + dt) \sim h(t) - f(t)dt$ and $h(0) = 0$ so that

$$g(c) = e^{h(c)} = e^{-\int_0^c f(t)dt} = (1 + Qc)^{-1/Q} \tag{14}$$

matching the previous results.

This approach has advantages and disadvantages. The main disadvantage is the difficulty of proving its validity. As the random process continues there will be more and more variance from expected behavior. It must be shown that the accumulated errors do not overwhelm the actual values. In essence we are dealing here with a stochastic differential equation. Indeed, proofs that the solution to this equation accurately portrays the situation look much like the original Rödl nibble. To examine the situation at $t = c$ we split the time interval into ce^{-1} intervals of some very small length ϵ. (Each time interval ϵ is a nibble.) With ϵ very small the solution to the corresponding difference equation is close to the solution of the differential equation. Suppose in time ϵ the expected change of a degree is some αD with α, ϵ comparable. Roughly speaking the variance in that change will go like $(\alpha D)^{1/2}$. We need $D\epsilon \gg 1$ so that the variance is small compared to the expected change.

A big advantage of this approach is that it can be extended past any finite time. Consider the Erdős-Hanani situation as a $(Q + 1)$-uniform hypergraph on $v = \Theta(n^k)$ vertices , regular of degree $D = \Theta(n^{k-l})$ with all codegrees $O(D/n)$. At finite time T the proportion of uncovered l-sets is $O(T^{-1/Q})$ for T large. Now suppose the differential equation can be shown to remain valid up to time n^γ for some positive constant γ. Then we get an improvement on Rödl's Theorem. The dynamic algorithm then gives a packing so that the proportion of uncovered l-sets is $O(n^{-\gamma'})$ for a calculable positive constant γ'. It isn't so easy – extending the range of validity of the differential equation to T a function of n requires great care with the errors introduced from all sources. However, this approach has been used with success by N. Wormald [21] and, independently, D. Grable [11].

A generalization of Pippenger's Theorem has been given by J. Kahn in unpublished work. Let H be a $(Q + 1)$-uniform hypergraph. Consider the following linear programming problem on real variables x_E, E ranging over the edges of H.

$$\text{maximize } \sum_{E \in H} x_E$$
$$\text{given } \sum_{v \in E} x_E \leq 1 \text{ for all } v \in V(H)$$
$$\text{and all } 0 \leq x_E \leq 1$$

A feasible solution to the above system is called a fractional packing. We let ν^* denote the solution to this linear programming problem. If we also require all $x_E \in \{0, 1\}$ this yields the packing $P = \{E : x_E = 1\}$ and the solution, denoted by ν, is the size of the maximal packing. Thus $\nu \leq \nu^*$.
Kahn's Theorem: For all Q and all $\epsilon > 0$ there exists $\delta > 0$ so that if x_E is a feasible solution to the above system with

$$\sum_{v,w \in E} x_E < \delta$$

for all distinct v, w then there is a packing P with

$$|P| \geq (1 - \epsilon) \sum_{E \in H} x_E$$

Roughly speaking, Kahn's Theorem says that under appropriate side conditions $\nu \sim \nu^*$. We shall indicate an argument for Kahn's Theorem by creating an appropriate continuous time process. We are given the hypergraph H and the values x_E. It will be convenient to set

$$y_v = \sum_{v \in E} x_E$$

so that all $y_v \in [0, 1]$. Give each E independently a birthdate t_E such that given E has not been born by t its probability of being born in the next infinitesmal time dt is $x_E dt$. Formally this is the exponential distribution, $\Pr[t_E > c] = e^{-cx_E}$.

As with Pippenger's theorem we dynamically keep a set $L = L_t$ and let S_t (the surviving edges) be the restriction of H to L_t. Again if $E \in S_t$ is born in infinitesmal time interval $[t, t + dt)$ it is added to the packing P so that $P_{t+dt} = P_t \cup \{E\}$. Now, however, we introduce the possibility of killing a vertex v. For each $v \in L_t$ define $y_v(t) = \sum_{v \in E \in S_t} x_E$. We think of this as the weighted degree of v at time t. Set $f(t) = (1 + Qt)^{-1}$ as before. Now we kill v in the time interval $[t, t + dt)$ with probability $(f(t) - y_v(t))dt$. (If this is negative v is not killed.) Killing v means v is removed from L so $L_{t+dt} = L_t - \{y\}$ and therefore all $E \in S_t$ containing y are no longer surviving. The claim now is that at time t most v have

$$y_v(t) \sim f(t) y_v$$

As $f(0) = 1$ this holds for $t = 0$, now assume it holds for t. Any $w \in L_t$ is part of a newly born (in $[t, t + dt)$) E with probability $y_w(t)dt$ and is killed with the compensating probability $(f(t) - y_w(t))dt$ so it is removed from L with probability $f(t)dt$. Given that v itself remains in L each of its edges E has probability $Qf(t)dt$ of having a vertex lost, which would subtract x_E from $\deg(v)$. Then the expected total loss in the weighted degree is $y_v(t)(Qf(t)dt) \sim Qy_v f^2(t)dt$. Since $f(t)$ satisfies 12 the expected new value of the weighted degree is $\sim y_v f(t + dt)$ as desired.

Certainly the above argument needs work to be made formally correct. But suppose its correctness and now consider a vertex v. Let $g(T)$ be the probability that $v \in L_T$. We get 14 as before so that $g(T) \to 0$. But in each infinitesmal time interval $[t, t + dt)$ the probabilities that v is in a newly born E and that v is killed off are in the ratio y_v to $1 - y_v$ – i.e., conditioning on $v \in L_{t+dt} - L_t$, $\Pr[v \in E \in P_{t+dt}] \sim y_v$. Thus $\Pr[v \in E \in P_T] \sim y_v(1 - g(T))$. Summing over all vertices v, the expected size

$$E[|\bigcup P_T|] \sim (1 - g(T)) \sum_v y_v = (Q + 1)(1 - g(T)) \sum_E x_E$$

so that as $T \to \infty$ the expected size of P_T approaches $\sum_E x_E$ as desired.

2.2   Ramsey $R(3, k)$

The Ramsey function $R(l, k)$ is defined as the minimal n so that any graph on n vertices must contain either a clique of size l or an independent set of size k. Existence of such n is Ramsey's Theorem itself. Asymptotics of the Ramsey function (and its numerous generalizations) have been closely linked with probabilistic methods from the beginning.

Theorem (Erdős (1947)[6]:

$$\binom{n}{k} 2^{-\binom{k}{2}} < 1 \Rightarrow R(k, k) > n \tag{15}$$

Proof. Take the random graph $G \sim G(n, p)$ with $p = \frac{1}{2}$. Then $\binom{n}{k} 2^{-\binom{k}{2}}$ is the expected number of cliques and independent sets of size k. When this number is less than one then with positive probability it is zero so that $R(k, k) > n$.

Here we concentrate on $l = 3$ and the asymptotics as $k \to \infty$. The basic upper bound, from the proof of Ramsey's Theorem, was $R(3, k) \leq \binom{k+1}{2}$ which was lowered to $O(k^2 \frac{\ln \ln k}{\ln k})$ by Graver and Yackel [12] in 1968 and then to $O(\frac{k^2}{\ln k})$ by Ajtai, Komlós and Szemerédi [1] in 1980. A lower bound $R(3, k) > n$ means that there exists a trianglefree graph G on n vertices with no independent k set. After a number of "false starts" a lower bound $R(3, k) = \Omega(\frac{k^2}{\ln^2 k})$ was shown by Erdős [7] in 1961. This paper displays a remarkable combination of insight and technical skill. Over the decades, as new techniques have emerged, a number of authors have reproven this result. My own effort [20] in 1977 used the Lovász Local Lemma. Perhaps the most elementary proof is due to Krivelevich[15], we repeat it here in essentially complete form. We use an elementary and quite useful lemma from [10].

Lemma: Let $A_1, \ldots, A_m$ be events with $\sum \Pr[A_i] = \mu$. Then the probability that there exist s of the events, say $A_{i_1}, \ldots, A_{i_s}$ which are mutually independent events and which all hold is at most $\mu^s / s!$. Proof. We bound $\sum \Pr[A_{i_1} \wedge \ldots \wedge A_{i_s}]$ over all such $i_1, \ldots, i_s$. With the A's mutually independent we replace this with $\sum \Pr[A_{i_1}] \cdots \Pr[A_{i_s}]$. This sum over all $i_1, \ldots, i_s \in \{1, \ldots, m\}$ is precisely μ^s and each desired term has been counted $s!$ times.

Krivelevich's Proof: Let $G \sim G(n, p)$ with $p = \epsilon n^{-1/2}$ and set $k = K n^{1/2} \ln n$ with $\epsilon = .1$ and $K = 10^6$ for definiteness though any moderately small ϵ and very large K would do. Let F be a (any) maximal family of edge disjoint triangles of G and let $G^* = G - \bigcup F$, i.e., G with all edges of all triangles of F removed. G^* is certainly trianglefree. For any k-set of vertices S the number X_S of edges of $G|_S$ has Binomial Distribution $B(\binom{k}{2}, p)$. Elementary large deviation results give

$$\Pr[X_S \leq \frac{1}{2} p \binom{k}{2}] < (0.9)^{\binom{k}{2} p}$$

and since (estimating $\binom{k}{2} \sim \frac{1}{2} k^2$ and $\binom{n}{k} \leq n^k$)

$$\binom{n}{k} (0.9)^{\binom{k}{2} p} < \left[n (0.9)^{pk/2} \right]^k \ll 1$$

almost surely *all* S have at least $\frac{1}{4} p k^2 \sim \frac{1}{4} \epsilon K^2 n^{1/2} \ln^2 n$ edges. Again fix S and consider all potential triangles efg (listing the edges) with $e \subset S$. For each let A_{efg} be the event that they all lie in G so that $\Pr[A_{efg}] = p^3$. There are $\binom{k}{2}(n-k) + \binom{k}{3} \sim \frac{1}{2}k^2 n$ such events so

$$\mu = \sum_{e \subset S} \Pr[A_{efg}] \sim \frac{1}{2}k^2 n p^3 = k\left(\frac{1}{2}\epsilon^3 K^2 \ln n\right)$$

Events A_{efg} are mutually independent when their edge sets are disjoint. From the Lemma above the probability that there are 3μ edge-disjoint triangles efg in G with $e \subset S$ is less than $\mu^{3\mu}/(3\mu)!$ and as

$$\binom{n}{k}\frac{\mu^{3\mu}}{(3\mu)!} < n^k(0.95)^{3\mu} < \left[n(0.95)^{\frac{3}{2}\epsilon^2 K^2 \ln n}\right] \ll 1$$

almost surely for *every* S there are less than 3μ such triangles and therefore $\bigcup F$ will have less than 9μ edges in S. Having picked ϵ, K so that $9\mu < \frac{1}{2}p\binom{k}{2}$ the elimination of these edges makes no S independent. Thus $R(3,k) > n$ or, reversing variables, $R(3,k) = \Omega(k^2/\ln^2 k)$.

We improve this classic result by thinking *dynamically*.

Consider the following random dynamic process to form a trianglefree graph G on vertices $1, \ldots, n$. To each pair $e = \{i,j\}$ assign a birthtime $x_e \in [0, n^{1/2}]$, independently and uniformly. At time zero G is empty. When e is born it is added to G if its addition does not create a triangle. We say e is accepted in that case, rejected otherwise. Let G_t denote G at time t. A pair $\{i,j\}$ is surviving at time t if it has not been born and there is no k with $\{i,k\}, \{j,k\}$ already in G - so that it would be accepted if born now. Let $S = S_t$ be the graph of surviving pairs. Let $g_n(c)$ be the probability that any particular e (they all look alike) is surviving at time t.

We define a continuous time branching process that will mirror the fate of e above. Begin at time c with a single "Eve". Split $[0, c] \times [0, c]$ into infinitesmal squares $[x, x + dx] \times [y, y + dy]$. With probability $dx \cdot dy$ Eve gives birth to twins with birthtimes x, y. Equivalently, Eve gives birth to X pairs of twins with X having Poisson distribution with mean c^2 and given the number of births all birthdates are independent (twins are not born at the same time) and uniform in $[0, c]$. A child born at time x then gives birth by the same process in $[0, x] \times [0, x]$. A rooted tree T is thus generated and it can be shown that with probability one T is finite. We call vertices of T surviving or dying as follows. All childless vertices are surviving. A vertex is dying if and only if it has a (at least one) birth where both twins are surviving. Working up from the leaves of the tree every vertex of T is so designated. Let $g(c)$ denote the probability that the root Eve survives.

We give a rough argument that $\lim_{n \to \infty} g_n(c) = g(c)$. For $e = \{v, w\}$ we look at those u for which $x_{vu} \le c$ and $x_{wu} \le c$. There are $n - 2$ potential u, each independently has this property with probability $(cn^{-1/2})^2$, so the number is asymptotically Poisson with mean c^2. Given that, the actual birthtimes are uniform in $[0, c]$. We then consider uv, uw twins of $e = vw$. We continue this process building up a tree. The analogy fails if some edge is child to two edges but this can be shown to occur with probability $o(1)$. Working backwards from the leaves one sees that an edge f is placed in G exactly when, considered as a vertex of T, it survives as described.

We find $g(c)$ by a differential equation. The difference $g(t) - g(t + dt)$ is the probability that Eve has no twins both born before t (probability $g(t)$) then has a pair of twins one of which is born in $[t, t + dt)$ (probability $2t \cdot dt$) and then they both survive. The twin born in $[t, t + dt)$ has probability $\sim g(t)$ to survive. The other is born uniformly in $[0, t]$ so its expected probability to survive is the average of $g(x)$ over the interval. This yields the differential equation

$$g'(t) = -2tg^2(t)\frac{1}{t}\int_0^t g(x)dx \tag{16}$$

or, setting $G(t) = \int_0^t g(x)dx$,

$$G''(t) = -2(G'(t))^2 G(t) \tag{17}$$

With initial conditions $G(0) = 0$ and $G'(0) = g(0) = 1$ this has a unique solution given implicitly by

$$t = \int_0^{G(t)} e^{x^2} dx \tag{18}$$

Here is a second approach to the same result. At a given time t let $\deg_S(i)$ denote the number of neighbors of vertex i in S, for $e = \{i, j\} \in S$ let $\deg_\Delta(e)$ denote the number of triangles containing e in S, and for $e = \{i, j\} \in S$ and designated i let $N(e, i)$ denote the number of k with $\{i, k\} \in S$ and $\{j, k\} \in G$. (In this case we call $\{i, j\}, \{i, k\}$ a cherry – if one is born the other dies.) Suppose that for every i

$$\deg_S(i) \sim a(t)n$$

and for every $e = \{i, j\} \in S$

$$\deg_\Delta(e) \sim b(t)n$$

$$N(e, i) \sim c(t)n^{1/2}$$

Now add an infinitesmal time dt and consider expectations. Each surviving e is in $\sim 2c(t)n^{1/2}$ cherries (half from each end) so with probability $2c(t)dt$ one of the other edges will be born and so e dies with probability $2c(t)dt$. Of the $a(t)n$ edges containing a given i an expected $2a(t)c(t)ndt$ die. Thus

$$a'(t) = -2a(t)c(t)$$

Similarly of the $b(t)n$ triangles containing e an expected $2b(t)n(2c(t)dt)$ will be "destroyed" in that one of their edges will die so

$$b'(t) = -4b(t)c(t)$$

Of the $c(t)n^{1/2}$ cherries containing e at i, $2c^2(t)n^{1/2}dt$ will be lost by having the other edge die but $b(t)n^{1/2}dt$ new cherries are created when an old triangle containing e has the edge not through i born. Thus

$$c'(t) = -2c^2(t) + b(t)$$

Further at time zero S is the complete graph so we have initial conditions $a(0) = 1 = b(0)$, $c(0) = 0$, yielding a unique solution. This has a nice solution in terms of $G(t)$ given by 18. Then

$$a(t) = e^{-G(t)^2} = G'(t) \qquad b(t) = a(t)^2, \qquad c(t) = G(t)a(t) \qquad (19)$$

At time t proportion $a(t)$ of the pairs are surviving so $\frac{1}{2}n^{1/2}a(t)dt$ pairs are accepted by time $t+dt$ so the expected total edges in G_t is $\frac{1}{2}n^{3/2}\int_0^t a(x)dx = \frac{1}{2}n^{3/2}G(t)$.

All this is lead in to our exciting finish. Jeong Han Kim [14] has found (up to constants) the asymptotics of $R(3, k)$. He has improved the Erdős lower bound to $R(3, k) > c\frac{k^2}{\ln k}$. Reversing parameters he shows the existence of a trianglefree graph G on n vertices with no independent set of size $k = Cn^{1/2}\ln^{1/2} n$. The method (at least from this author's vantagepoint) is to consider dynamically the random trianglefree graph as described above. At time t it has $\sim \frac{1}{2}n^{3/2}G(t)$ edges. From 18 one sees $G(t) \sim \ln^{1/2} t$ as $t \to \infty$. Kim shows that the solutions $a(t), b(t), c(t)$ given above remain asymptotically valid up to $t = n^\gamma$, for γ a small but absolute constant. To be sure, this is considerably more difficult then showing validity for a finite time interval. The first method is not strong enough for this, he uses (basically) the second approach. Keeping bounds on the error terms brought on by the randomness requires mastery of the martingale inequalities. At $t = n^\gamma$ some $cn^{3/2}\ln^{1/2} n$ edges have been accepted to G. A random graph with this many edges has no independent set of size $k = Cn^{1/2}\ln^{1/2} n$. To be sure, this graph is anything but random. Still Kim shows that for any k-set S the probability that S remains independent is basically what it would be were G random. This yields the solution to a sixty year old problem, the asymptotics of $R(3, k)$.

References

[1] M. Ajtai, J. Komlós, E. Szemerédi, A note on Ramsey numbers, *J. Combinatorial Theory (Ser A)*, 29 (1980), 354-360

[2] N. Alon, J. Spencer, *The Probabilistic Method*, John Wiley, New York, 1991

[3] B. Bollobás, *Random Graphs*, Academic Press, London, 1985

[4] B. Bollobás, The chromatic number of random graphs, *Combinatorica* 8 (1988), 49-55

[5] R. Boppana, J. Spencer, A Useful Elementary Correlation Inequality, *J. Combinatorial Theory - Ser. A* 50 (1989), 305-307

[6] P. Erdős, Some remarks on the theory of graphs, *Bull. Amer. Math. Soc.* 53 (1947), 292-294

[7] P. Erdős, Graph Theory and Probability II., *Canad. J. Math.* 13 (1961), 346-352

[8] P. Erdős and H. Hanani, On a limit theorem in combinatorial analysis, *Publ. Math. Debrecen*, 10 (1963), 10-13

[9] P. Erdős and A. Rényi. On the evolution of random graphs. *Magyar Tud. Akad. Mat. Kut. Int. Közl* 5 (1960), 17-61

[10] P. Erdős, P. Tetali, Representations of integers as the sum of k terms, *Random Structures & Algorithms*, 1 (1990), 245-261

[11] D. Grable, More-Than-Nearly-Perfect Packings and Partial Designs, (preprint 1994)

[12] J.E. Graver, J. Yackel, Some graph theoretic results associated with Ramsey's theorem. *J. Combinatorial Theory* 4 (1968), 125-175

[13] S. Janson, T. Luczak and A. Rucinski. An exponential bound for the probability of nonexistence of specified subgraphs of a random graph, in *Procedings of Random Graphs '87*, M. Karonski et. al. eds, J. Wiley 1990, 73-87

[14] J.H. Kim, The Ramsey Number $R(3,t)$ has Order of Magnitude $t^2/\log t$, (preprint, 1994)

[15] M. Krivelevich, Bounding Ramsey numbers through large deviation inequalities, (preprint, 1994)

[16] C. McDiarmid, On the method of bounded differences, in J. Siemons, ed., *Surveys in Combinatorics 1988*, Cambridge Univ. Pr. 1989, 148-188

[17] N. Pippenger, J. Spencer, Asymptotic behavior of the chromatic number for hypergraphs, *J. Combinatorial Theory (Ser. A)*, 51 (1989), 24-42

[18] V. Rödl, On a packing and covering problem, *European Journal of Combinatorics*, 5 (1985), 69-78

[19] E. Shamir, J. Spencer, Sharp concentration of the chromatic number on random graphs $G_{n,p}$. *Combinatorica* 7 (1987), 121-129

[20] J. Spencer, Asymptotic Lower Bounds for Ramsey Functions, *Discrete Math* 20 (1977), 69-76

[21] N. Wormald, Differential Equations for Random Processes and Random Graphs, (preprint 1994)

Author Address:
Courant Institute
251 Mercer St.
New York, NY 10012
USA
spencer@cs.nyu.edu

Printed in the United States
By Bookmasters